W9-BVQ-640

A PROGRAMMED
INTRODUCTION
TO STATISTICS

SECOND EDITION

A PROGRAMMED INTRODUCTION TO STATISTICS

SECOND EDITION

Freeman F. Elzey

BROOKS/COLE PUBLISHING COMPANY

Pacific Grove, California

L.C. Cat. Card No.: 79-161489
ISBN 0-8185-0018-2
Printed in the United States of America

20 19 18 17

PREFACE

Like the preceding edition, this programmed introduction to statistics presents instruction in a logical sequence that allows the student to participate actively in the instruction process. It is written for students being introduced for the first time to statistical techniques and to the application of those techniques. It may be used for individual study or in undergraduate or graduate courses. It may be used as the only statistical text for a course or as an auxiliary text.

This edition differs from the first edition in a number of ways. It incorporates additions and changes in content suggested by instructors who have used the first edition in their classes, and it includes the correction of errors that are invariably present in first editions of statistical texts. The sets have been reorganized in a more logical order so that the sequential development of material more clearly parallels that presented in the major statistical texts. The measures of variability have been reorganized so that the presentation of variance and standard deviation immediately follows the discussion of range and average deviation. In this edition the discussion of the normal distribution leads logically to inferential statistics. Some changes in symbolic notation have been made to reflect current statistical practice.

The presentation of graphic methods has been expanded and a new programmed set has been added. There is a completely revised exposition on the use of confidence intervals. The distinction between sample statistics and population parameters as well as their symbolic notations has been clarified.

In this edition, all formulas are presented in the text as well as in the Appendix. Additional space has been provided for all student responses. A square root table and instructions for its use has been added.

This text is directed primarily to the student who is unfamiliar either with the basic concepts of statistical techniques or with the mathematics needed to apply these techniques. Only a rudimentary knowledge of algebra is needed. This program is a beginning course, and stresses application; it does not attempt to develop theoretical or mathematical derivations of the various techniques.

Statistics is a difficult subject for many students. The major reasons for this may be that too much instruction is given at one time, that the material is not logically organized so that the student can follow its development, or that the student is not actively engaged in the instructional procedure.

Programmed instruction attempts to solve these problems by allowing the student to participate actively in every portion of the instructional process, by presenting instruction organized so that each step leads in logical sequence to the next, and by allowing the student to proceed at his own pace through the program—quickly through those areas that present no difficulty and more slowly where he feels it necessary. An additional advantage in programmed instruction is that after each finite step in the program, the student is informed immediately of the accuracy of his understanding. This immediate feedback is one of the important features of programmed instruction. It provides correction when necessary; more important, it verifies when the material has been correctly grasped, thus reinforcing learning.

The first edition of this material was thoroughly field tested. Copies of the original manuscript were used by hundreds of students in several types of courses involving the study of statistics. The rate of student error in frame responses was well under five percent, and before-and-after testing reflected gratifying increases in student mastery of statistics.

This text is logically organized into twenty-five sets. Each set is self-contained and can usually be completed at one sitting. A brief set introduction describes the contents of each set and presents specific objectives for that set. These stated objectives alert the student to the important aspects of statistics to be covered in that set.

Each set contains a series of frames written and sequenced so that the required responses can be determined easily. The correct response is given immediately below each frame. It is this process of presenting the material in sequenced frames, each requiring a positive response, then immediately providing a check for the accuracy of the response, that constitutes the concept of "programmed" instruction.

At the end of each set is a series of exercises. They are an integral part of the instruction, providing a self-test and giving the student opportunity to apply what he has learned. The exercises parallel exactly the objectives set out in the introduction to each set.

Another unique feature of this book is the presentation of formulas, tables, and a glossary of symbols at the rear of the book. These can be removed for convenient reference while the student is working in this text and kept for permanent reference later.

All data presented in this text are fictitious and were developed specifically to illustrate this program. In order to reduce computational drudgery, the amount of data presented is kept to the minimum necessary to illustrate the statistical techniques.

Conventional symbolic notation has been used throughout the text so that it may be used with standard statistical texts without confusion.

I wish to express my sincere appreciation to Professor Samuel Levine for his advice and encouragement during the writing of this text. Grateful acknowledgement is also given to Professors Derek Nunney of Wayne State University, Gerald C. Helmstadter of Arizona State University, Frederick J. McDonald of Stanford University, Omer J. Rupiper of the University of Oklahoma, Garlie A. Forehand of Carnegie Institute of Technology, and Harold Jonsson and Samuel Levine of San Francisco State College for field testing preliminary editions of this text in their classes.

My particular gratitude goes to Bert Baecher for generously making available the appropriate psychological surroundings during the writing of this book.

I am indebted to the literary executor of the late Sir Ronald A. Fisher, F.R.S., Cambridge, to Dr. Frank Yates, F.R.S., Rothamsted, and to Messrs Oliver & Boyd Limited, Edinburgh, for permission to reprint Tables Nos. III and IV from their book *Statistical Tables for Biological, Agricultural and Medical Research.*

I am also indebted to the literary executor of the late Sir Ronald A. Fisher, F.R.S., Cambridge and to Oliver & Boyd Limited, Edinburgh, for their permission to reprint Table No. V.A. from their book *Statistical Methods for Research Workers.*

While only the author can assume responsibility for the content of a book, and for any errors which may have escaped, I would nevertheless like to express my thanks to Andrew Watson, whose editorial ability contributed measurably to the first edition of this book, and to Terry Hendrix, who organized and directed the field testing.

CONTENTS

A PROGRAMMED INTRODUCTION TO STATISTICS

SECOND EDITION

INTRODUCTION

In this program you will learn a number of statistical techniques, their statistical formulas, and when and how to use them. In order to use this program, you need only a rudimentary knowledge of algebraic procedures (that is, the use of symbols and the solving of simple algebraic equations). Even if you are weak in these areas, you should be able to use this program, since much help in computation is given, especially in the early sets.

WHY THIS TEXT IS IN PROGRAMMED FORM

This text presents a new method for learning statistics in which you: (1) actively participate in every portion of the instructional process; (2) are presented with step-by-step instruction organized so that each step leads logically to the next; and (3) are allowed to proceed at your own pace through the program, moving quickly through those areas that present no difficulty and more slowly where you feel it necessary. After each finite step in the program, you are immediately informed of the correctness of your understanding. This immediate feedback is one of the important features of programmed instruction. It provides correction when necessary; but more important, it verifies when you have correctly grasped what is being taught.

HOW TO USE THIS PROGRAMMED TEXT

This program is divided into twenty-five sets. Each set contains three components: (1) an introduction to the set, (2) the programmed instruction portion of the set, and (3) a series of exercises.

(1) The Introduction

Before each set in the program a brief introductory statement describes the content of the programmed portion of the set and provides additional information not contained in the programmed portion. You should read the introduction to each set carefully since it gives you a frame of reference for the instruction that follows. The introduction also states the specific objectives to be achieved in the set. These objectives alert you to the important aspects of statistics to be presented in the set and provide a clear picture of what you can expect to learn from the set.

(2) The Program

Each set of the program contains a number of small units, called *frames*. Each frame presents some information and includes a blank space which you are to fill in. The correct response to each frame, which is given immediately below it, should be kept covered with a card or sheet of paper until after you have written your response to the frame.

You are given some cues as to the type of response required in each frame. For instance, the number and size of the blanks indicate the number and size of words required. In some frames there is a series of alternatives from which you are to choose. For example, "Grass is _____ (green/red/blue)." Some frames require you to provide a symbol. This is indicated by the word (symbol) following the blank. For example, "When you wish to express dollars you use __ (symbol)."

(3) The Exercises

At the conclusion of each set there is a series of exercises. The exercises in this book are an integral part of the instructional process and all are to be completed. They provide a self-test by which you may determine whether you have grasped the material in the set. You will notice that the exercises exactly parallel the objectives as stated in the introduction to each set. The additional experience of performing these exercises increases and reinforces your understanding of what you have just learned. The answers are provided in the rear of the text.

Formulas, Tables, and a Glossary of Statistical Symbols are presented at the rear of the text. These will frequently be referred to as you proceed through the frames of this program. They are perforated and should be removed and examined when referred to in the text. The Formulas, Tables, and Glossary provide you with a permanent reference for further statistical work.

The data presented in this program are fictitious and are kept to a minimum to reduce the amount of computation required. It should be pointed out that the usual research study includes much more data than is given in the illustrations.

A WORD OF CAUTION

For this programmed text to be most effective, you must follow the instructions above explicitly. Any shortcut reduces your chance of learning and retaining the statistical concepts and procedures covered. You should:

(1) Read the introduction and objectives for each set.

(2) Read each frame carefully, writing in your response to each blank provided in the frame. Keep the correct answer below the frame *covered* until you have written in your response.

(3) After responding, check your response with the correct answer below the frame. If your response is correct, proceed to the next frame. If your answer is incorrect, you should review the preceding frames in order to find out why you were incorrect. Make the necessary correction in your response before proceeding.

(4) Do all the exercises at the end of each set. Check your answers with those provided at the rear of the text.

By following these instructions, you will be assured of getting the most out of this programmed text in beginning statistics.

ORGANIZATION OF DATA:
FREQUENCY DISTRIBUTIONS

Statistics is the study concerned with the analysis of observed facts which have been expressed as numbers. These numbers may be test scores, linear measurements, frequencies of occurrence, numbers of people, and so on. When a researcher obtains a quantity of these numbers in order to describe or make an inference about a phenomenon, they are usually called *data.*

How the researcher *obtains* these data falls in the realm of research design; how he *examines* these data falls in the realm of statistics. In this book we shall be concerned only with the analysis of statistical data, not in how they were obtained.

We will be concerned initially with procedures for describing data. Later we will examine some statistical techniques to let us draw inferences from these data.

The first step in the statistical analysis of a mass of numerical data is to arrange the data in some order. This set presents definitions of the terms *data, raw score, variability,* and *frequency,* and illustrates the preparation of frequency distributions of both ungrouped and grouped data, and the determination of the upper and lower limits for a class interval.

SPECIFIC OBJECTIVES OF SET 1

At the conclusion of this set you will be able to:

(1) prepare a frequency distribution of ungrouped data.

(2) group data into class intervals.

(3) prepare a frequency distribution of grouped data.

(4) identify the real lower and upper limits of any particular raw score.

(5) identify the symbols X, f, N, i, $\ell\ell$

1. In education and psychology, measurements of certain characteristics are obtained for groups of individuals. When you _____ the intelligence, height, or weight of a group of individuals, you obtain a set of numbers.

■ ■ ■ ■ ■ ■ ■ ■ ■ ■ ■ ■ ■ ■

measure

---

2. The technical term for a set of numbers is *data*. A set of numbers indicating the individual arithmetic scores for a group of children is called _____.

■ ■ ■ ■ ■ ■ ■ ■ ■ ■ ■ ■ ■ ■

data

---

3. The term *raw score* is used to indicate the measurement, or datum, obtained for one individual. Thus, the number of items a person passes on an intelligence test is termed a _____ _____.

■ ■ ■ ■ ■ ■ ■ ■ ■ ■ ■ ■ ■ ■

raw score

---

4. If John Smith receives 95 points in an arithmetic test, he is said to have a _____ _____ of 95 in arithmetic.

■ ■ ■ ■ ■ ■ ■ ■ ■ ■ ■ ■ ■ ■

raw score

---

5. Because people vary in their ability to do arithmetic, it is likely that the raw scores of the group also _____.

■ ■ ■ ■ ■ ■ ■ ■ ■ ■ ■ ■ ■ ■

vary

---

6. When you collect data on a number of individuals, you find that the _____ _____ for all individuals are not the same.

■ ■ ■ ■ ■ ■ ■ ■ ■ ■ ■ ■ ■ ■

raw scores

Plate 1. Spelling Scores

10
12
9
12
7
8
12

7. This plate gives the spelling scores that seven people received on a spelling test. These spelling scores may be called _____ _____ .

■ ■ ■ ■ ■ ■ ■ ■ ■ ■ ■ ■ ■ ■

raw scores

8. PLATE 1. To describe this set of raw scores best, you first need to *order* them. To do this, you arrange the raw scores in numerical order with the largest score at the top. Order the raw scores in Plate 1.

■ ■ ■ ■ ■ ■ ■ ■ ■ ■ ■ ■ ■ ■

Plate 2. Spelling Scores

12
12
12
10
9
8
7

9. PLATE 2. The highest score value is 12 and the lowest score value is ___.

■ ■ ■ ■ ■ ■ ■ ■ ■ ■ ■ ■ ■ ■

7

10. PLATE 2. The largest score value is 12 and the smallest score value is 7. Therefore, a raw score for any particular individual in this group must be somewhere between ___ and ___.

■ ■ ■ ■ ■ ■ ■ ■ ■ ■ ■ ■ ■ ■

7, 12

---

11. If the largest score value is 12 and the smallest score value is 7, it is possible for an individual to receive any one of six different score values. The possible score values are ___ , ___ , ___ , ___ , ___ , and ___ .

■ ■ ■ ■ ■ ■ ■ ■ ■ ■ ■ ■ ■ ■

12, 11, 10, 9, 8, 7

---

12. The number of times a raw score occurs in a set of data is called the *frequency* of that score. In plate 2, the frequency of score value 12 is 3, because this raw score occurs _____ times.

■ ■ ■ ■ ■ ■ ■ ■ ■ ■ ■ ■ ■ ■

three

---

13. When you list the frequency with which each score value occurs in a set of raw scores, you have a *frequency distribution* of the ___ _____ .

■ ■ ■ ■ ■ ■ ■ ■ ■ ■ ■ ■ ■ ■

raw scores

Plate 3. Frequency Distribution of Intelligence Test Scores

X	f
105	1
104	2
103	3
102	4
101	3
100	0
99	2
98	1
	$N = 16$

14. This plate gives the _____ _____ of a set of raw scores.

■ ■ ■ ■ ■ ■ ■ ■ ■ ■ ■ ■ ■ ■

frequency distribution

---

15. PLATE 3. Each possible score value is identified in the X column, and the frequency of each score value is listed in the __ (symbol) column.

■ ■ ■ ■ ■ ■ ■ ■ ■ ■ ■ ■ ■ ■

f

---

16. PLATE 3. The symbol used for the score value is __ (symbol).

■ ■ ■ ■ ■ ■ ■ ■ ■ ■ ■ ■ ■ ■

X

---

17. PLATE 3. The frequency for score value 99 is __.

■ ■ ■ ■ ■ ■ ■ ■ ■ ■ ■ ■ ■ ■

2

---

18. PLATE 3. From this plate you learn that the symbol f represents the _____ of each score value in the frequency distribution.

■ ■ ■ ■ ■ ■ ■ ■ ■ ■ ■ ■ ■ ■

frequency

---

19. PLATE 3. For the score value of 100, the f is zero because _____ subjects have the raw score of 100.

■ ■ ■ ■ ■ ■ ■ ■ ■ ■ ■ ■ ■ ■

no

9

20. **PLATE** 3. From this plate you learn that N is used to represent the total number of ____ _____ in the frequency distribution.

■ ■ ■ ■ ■ ■ ■ ■ ■ ■ ■ ■ ■ ■

raw scores

---

21. Prepare the frequency distribution for the raw scores presented in Plate 2. (Use plate 3 as a guide.)

X	f

■ ■ ■ ■ ■ ■ ■ ■ ■ ■ ■ ■ ■ ■

Plate 4. Frequency Distribution of Spelling Scores

X	f
12	3
11	0
10	1
9	1
8	1
7	$\underline{1}$
	$N = \underline{}$

---

22. PLATE 4. $N = $____.

■ ■ ■ ■ ■ ■ ■ ■ ■ ■ ■ ■ ■ ■

7

---

23. PLATE 4. $N = 7$ because 7 is the total_____of individuals for which there are spelling scores.

■ ■ ■ ■ ■ ■ ■ ■ ■ ■ ■ ■ ■ ■

number

---

24. PLATE 4. For the score value of 11, the f is zero because there are____ individuals having the raw score of 11.

■ ■ ■ ■ ■ ■ ■ ■ ■ ■ ■ ■ ■ ■

no

---

25. PLATE 4. For the score value of 12, $f = 3$ because there are_____ individuals having the raw score of 12.

■ ■ ■ ■ ■ ■ ■ ■ ■ ■ ■ ■ ■ ■

three

---

26. In a frequency distribution, each score value is listed in the____(symbol) column and the frequency of each score is listed in the__(symbol) column.

■ ■ ■ ■ ■ ■ ■ ■ ■ ■ ■ ■ ■ ■

X, f

When the frequency distribution presents the f for each separate score value, the distribution is for *ungrouped* data.

Plate 5. Frequency Distribution for Ungrouped Data

X	f	X	f	X	f
30	1	23	0	16	1
29	1	22	2	15	1
28	0	21	2	14	0
27	1	20	1	13	0
26	2	19	3	12	0
25	1	18	0	11	1
24	0	17	1	10	1
				$N =$	19

27. This plate gives the frequency distribution for _____ data.

■ ■ ■ ■ ■ ■ ■ ■ ■ ■ ■ ■ ■ ■

ungrouped

---

28. PLATE 5. This plate gives the frequency distribution for ungrouped data because the f for each _____value is listed separately.

■ ■ ■ ■ ■ ■ ■ ■ ■ ■ ■ ■ ■ ■

score

---

29. Frequencies are sometimes presented for *groups* of score values. A distribution in which the score values are grouped is called a frequency distribution for _____ data.

■ ■ ■ ■ ■ ■ ■ ■ ■ ■ ■ ■ ■ ■

grouped

---

30. When a large range of score values is involved in a frequency distribution, it becomes desirable to _____the score values into what are called *class intervals*.

■ ■ ■ ■ ■ ■ ■ ■ ■ ■ ■ ■ ■ ■

group

---

31. PLATE 5. These data can be _____ into *class intervals* encompassing, for example, three score values each.

■ ■ ■ ■ ■ ■ ■ ■ ■ ■ ■ ■ ■ ■

grouped

Plate 6. Frequency Distribution for Grouped Data

Interval	f
28-30	2
25-27	4
22-24	2
19-21	6
16-18	2
13-15	1
10-12	2
	$N = 19$

32. This plate presents a frequency distribution for _____ data. This distribution is for the same data that were presented in Plate 5 as a distribution for ungrouped data.

■ ■ ■ ■ ■ ■ ■ ■ ■ ■ ■ ■ ■ ■

grouped

---

33. PLATE 5. The frequency of the three score values of 10, 11, and 12 are:

X	f	
12	0	The sum of the f's for these
11	1	three score values is ___.
10	1	

■ ■ ■ ■ ■ ■ ■ ■ ■ ■ ■ ■ ■ ■

2

---

34. PLATE 5. When the f's of these three score values are grouped into the class interval 10-12, the f for this interval is ___.

X	f	Interval	f
12	0		
11	1	10-12	2
10	1		

■ ■ ■ ■ ■ ■ ■ ■ ■ ■ ■ ■ ■ ■

2

---

35. The symbol i equals the number of values represented in the class interval. When the f's for the three score values of 10, 11, and 12 are grouped for the class interval 10-12, $i =$ ___.

■ ■ ■ ■ ■ ■ ■ ■ ■ ■ ■ ■ ■ ■

3

---

36. For the class interval 11-15, $i =$ ___.

■ ■ ■ ■ ■ ■ ■ ■ ■ ■ ■ ■ ■ ■

5

---

37. For class interval 11-15, $i = 5$ because there are _____ score values encompassed in the interval.

■ ■ ■ ■ ■ ■ ■ ■ ■ ■ ■ ■ ■ ■

five

---

38. The five score values encompassed in the class interval 11-15 are _____, _____, _____, _____, and _____.

■ ■ ■ ■ ■ ■ ■ ■ ■ ■ ■ ■ ■ ■

11, 12, 13, 14, 15

---

39. Determine the f for the class interval 12-14.

Ungrouped data		Grouped data	
X	f	Interval	f
14	3		
13	2	12-14	____
12	3		

■ ■ ■ ■ ■ ■ ■ ■ ■ ■ ■ ■ ■

8

40. PLATE 5. Group the data by class intervals containing three score values each, beginning with i 10-12, and prepare the frequency distribution.

Interval	f

Plate 6. Frequency Distribution for Grouped Data

Interval	f
28-30	2
25-27	4
22-24	2
19-21	6
16-18	2
13-15	1
10-12	2
	$N = 19$

41. Frequency distributions for grouped data are usually presented when there are _____ (few/many) different score values in the distribution.

■ ■ ■ ■ ■ ■ ■ ■ ■ ■ ■ ■ ■ ■

many

Note: For simplicity of presentation, the examples here are for small groups with few score values involved. Typically, you do not group the data when you have less than fifteen or twenty different score values in the distribution.

42. The *real limits* of each score value lie .5 points above and below the score value. Thus, for score value 10, the real upper limit is 10.5 and the real lower limit is _____.

■ ■ ■ ■ ■ ■ ■ ■ ■ ■ ■ ■ ■ ■

9.5

---

43. 11.5 and 12.5 are the *real limits* of the score value _____.

■ ■ ■ ■ ■ ■ ■ ■ ■ ■ ■ ■ ■ ■

12

---

44. The symbol for the *lower limit* of a score value is $\ell\ell$. The $\ell\ell$ for the score value 16 is _____.

■ ■ ■ ■ ■ ■ ■ ■ ■ ■ ■ ■ ■ ■

15.5

---

45. 9.5 is the _____ (symbol) of the score value 10.

■ ■ ■ ■ ■ ■ ■ ■ ■ ■ ■ ■ ■ ■

$\ell\ell$

---

46. For frequency distributions of grouped data, the real lower limit ($\ell\ell$) of a class interval is _____ score points below the lowest score value encompassed in the interval.

■ ■ ■ ■ ■ ■ ■ ■ ■ ■ ■ ■ ■ ■

.5

---

47. In addition to representing the lower limits of a single score value, the symbol $\ell\ell$ also represents the lower limits of a class interval. 7.5 is the _____ (symbol) of the class interval 8-10.

■ ■ ■ ■ ■ ■ ■ ■ ■ ■ ■ ■ ■ ■

$\ell\ell$

---

48. The real upper limit of interval 8-10 is _____ .

■ ■ ■ ■ ■ ■ ■ ■ ■ ■ ■ ■ ■ ■ ■

10.5

---

49. PLATE 6. The $\ell\ell$ of the top-most interval is _____ .

■ ■ ■ ■ ■ ■ ■ ■ ■ ■ ■ ■ ■ ■ ■

27.5

---

50. The real lower and upper limits of interval 25-27 are _____ and _____ .

■ ■ ■ ■ ■ ■ ■ ■ ■ ■ ■ ■ ■ ■ ■

24.5, 27.5

---

51. PLATE 6. The real lower and upper limits of the entire frequency distribution is _____ and _____ .

■ ■ ■ ■ ■ ■ ■ ■ ■ ■ ■ ■ ■ ■

9.5, 30.5

Plate 7.

Interval	f
26-30	3
21-25	4
16-20	6
11-15	3
6-10	3
1-5	2
	N = 21

52. In Plate 6, i was equal to three score points because each class interval encompasses three score values. In Plate 7, $i =$ ___ .

■ ■ ■ ■ ■ ■ ■ ■ ■ ■ ■ ■ ■ ■

5

53. PLATE 7. The $i = 5$ because there are _____ score values encompassed in each interval.

■ ■ ■ ■ ■ ■ ■ ■ ■ ■ ■ ■ ■

five

---

54. The value of i for a frequency distribution containing the interval 50-59 is ___ .

■ ■ ■ ■ ■ ■ ■ ■ ■ ■ ■ ■ ■

10

---

1. Prepare an ungrouped frequency distribution for the following set of raw scores. What is the N for this distribution?

10	6	15	8
11	11	12	9
8	9	7	1
5	2	5	

2. Prepare a frequency distribution for the set of raw scores above, grouping them into class intervals with $i = 3$.

3. Prepare a frequency distribution for the following set of raw scores. Let $i = 4$. What is the N of this distribution?

31	33	23
29	44	30
35	33	36
40	21	34
38	37	35
38	41	

4. Determine the real lower limits for the following raw scores.

9	107	1	36

5. Determine the real lower limits for the following class intervals.

11-15	1-3	4-9

6. What do the following symbols represent?

X	i
f	$\ell\ell$
N	

MEASURES OF CENTRAL TENDENCY: MODE AND MEDIAN

Data that have been collected and organized into a frequency distribution are ready for statistical analysis. The first step in the analysis of these data is to determine the one score value that best characterizes the entire frequency distribution. In statistics this value is called a measure of central tendency because it is the central value around which the scores tend to cluster. This set presents two measures of central tendency: the *mode* and the *median*; methods for determining these figures for ungrouped and for grouped data are also presented.

SPECIFIC OBJECTIVES OF SET 2

At the conclusion of this set you will be able to:

(1)   define the term *mode*.
(2)   define the term *median*.
(3)   locate the mode of a frequency distribution of ungrouped data.
(4)   locate the mode of a frequency distribution of grouped data.
(5)   calculate the median for grouped data using Formula 1.
(6)   calculate the median for ungrouped data using Formula 1.
(7)   identify the symbols Mdn, Σ, Σf_b, f_w.

1. When you have a large number of raw scores in a frequency distribution, they tend to group at some central or representative score. This central score is called a *measure of central tendency*. One measure of _____ _____ is called the mode.

■ ■ ■ ■ ■ ■ ■ ■ ■ ■ ■ ■ ■ ■

central tendency

---

2. The *mode* is defined as the most recurring or most frequent score in a frequency distribution.

Plate 8.

X	f
12	3
11	0
10	1
9	1
8	1
7	1
	$N = 7$

The modal score or *mode* for this frequency distribution is score _____.

■ ■ ■ ■ ■ ■ ■ ■ ■ ■ ■ ■ ■ ■

12

---

3. PLATE 8. The score value 12 is the mode of the distribution because it is the most _____ score in the distribution.

■ ■ ■ ■ ■ ■ ■ ■ ■ ■ ■ ■ ■ ■

frequent

---

4. PLATE 3, page 8. The most recurring or most frequent score for this frequency distribution is score _____. This score is the _____ of the distribution.

■ ■ ■ ■ ■ ■ ■ ■ ■ ■ ■ ■ ■ ■

102, mode

---

5. Since it is a measure of central tendency, the mode tells you nothing about the range of raw scores or their variability in the distribution. It only tells you which raw score occurs most _____.

■ ■ ■ ■ ■ ■ ■ ■ ■ ■ ■ ■ ■ ■

frequently

---

6. For grouped data, the interval that contains the largest *f* is called the *modal* interval of that distribution.

PLATE 6, page 13. The interval _____ - _____ is the modal interval of the distribution.

■ ■ ■ ■ ■ ■ ■ ■ ■ ■ ■ ■ ■ ■

19-21

---

7. The interval 19-21 is the modal interval of the distribution because it is the interval that contains the _____ *f* of scores.

■ ■ ■ ■ ■ ■ ■ ■ ■ ■ ■ ■ ■ ■

largest

---

8. PLATE 7, page 17. The modal interval of the distribution is interval _____ - _____.

■ ■ ■ ■ ■ ■ ■ ■ ■ ■ ■ ■ ■ ■

16-20

---

9. Another measure of central tendency is the *median*. It is defined as the midpoint (center) of the distribution. Thus, it is the point above which and below which ____ % of the raw scores lie.

■ ■ ■ ■ ■ ■ ■ ■ ■ ■ ■ ■ ■ ■

50

10. The median differs from the mode in that the median is the
    _____ of the distribution, whereas the mode is the score value or
    class interval that has the largest f.

    ■ ■ ■ ■ ■ ■ ■ ■ ■ ■ ■ ■ ■ ■

    midpoint

---

11. The median of a distribution in which there are seven scores will be the
    fourth score because_____ scores lie above and_____scores lie
    below this score.

    ■ ■ ■ ■ ■ ■ ■ ■ ■ ■ ■ ■ ■ ■

    three, three

---

12. PLATE 8. The symbol *Mdn* stands for the median. The score 10 is the
    _____ (symbol) of the distribution.

    ■ ■ ■ ■ ■ ■ ■ ■ ■ ■ ■ ■ ■ ■

    Mdn

---

13. PLATE 8. Score 10 is the *Mdn* because it is the_____ score in the
    frequency distribution.

    ■ ■ ■ ■ ■ ■ ■ ■ ■ ■ ■ ■ ■ ■

    middle

---

14. PLATE 3, page 8. The score 102 is both the_____and the
    _____ of the frequency distribution.

    ■ ■ ■ ■ ■ ■ ■ ■ ■ ■ ■ ■ ■ ■

    mode, median

Plate 9.

X	f
90	3
89	3
88	1
87	4
86	1
85	1
	$N = 13$

15. The mode of the frequency distribution is_____. The *Mdn* is_____.

■ ■ ■ ■ ■ ■ ■ ■ ■ ■ ■ ■ ■

87, 88

---

16. PLATE 9. Score 87 is the_____ of the distribution because it is the most_____ score.

■ ■ ■ ■ ■ ■ ■ ■ ■ ■ ■ ■ ■

mode, recurring

---

17. PLATE 9. Score 88 is the_____ (symbol) of the distribution because six raw scores are above and six raw scores are below it.

■ ■ ■ ■ ■ ■ ■ ■ ■ ■ ■ ■ ■

Mdn

Formula 1. Calculation of the Median

$$Mdn = \ell\ell + \left(\frac{.5N - \Sigma f_b}{f_w} \right) i$$

in which Σ = the sum of
f_b = frequency below the interval which contains the *Mdn*
f_w = frequency within the interval which contains the *Mdn*

---

18. Formula 1 presents the formula for the *Mdn* and the definitions of the new symbols. One of the new symbols introduced is the capital Greek letter Σ, which is pronounced "sigma" and means ___ _____ ___ .

■ ■ ■ ■ ■ ■ ■ ■ ■ ■ ■ ■ ■

the sum of

Note: If you have not already done so, remove the Formulas and Tables from the rear of the text and refer to them as directed.

19. FORMULA 1. The symbol $\ell\ell$ means the_____ _____ of the class interval which contains the *Mdn*.

■ ■ ■ ■ ■ ■ ■ ■ ■ ■ ■ ■ ■

lower limit

---

20. FORMULA 1. The symbol Σf_b means the_____ of the frequencies below the class interval which contains the *Mdn*.

■ ■ ■ ■ ■ ■ ■ ■ ■ ■ ■ ■ ■

sum

---

21. FORMULA 1. The symbol f_w equals the _____ within the class interval which contains the *Mdn*.

■ ■ ■ ■ ■ ■ ■ ■ ■ ■ ■ ■ ■

frequency

---

22. FORMULA 1. The symbol *i* represents the _____ of score values encompassed by each class interval in the distribution.

■ ■ ■ ■ ■ ■ ■ ■ ■ ■ ■ ■ ■

number

---

23. PLATE 6, page 13. The interval = ___ because there are three score values encompassed by each class interval.

■ ■ ■ ■ ■ ■ ■ ■ ■ ■ ■ ■ ■ ■

3

---

24. FORMULA 1. .5N means that you multiply the value of N by ___ .

■ ■ ■ ■ ■ ■ ■ ■ ■ ■ ■ ■ ■ ■

.5

---

25. .5 is the decimal form of ___ %.

■ ■ ■ ■ ■ ■ ■ ■ ■ ■ ■ ■ ■ ■

50

---

26. .5N is used in the formula for the *Mdn* because you wish to determine the score above which and below which ___ % of the raw scores lie.

■ ■ ■ ■ ■ ■ ■ ■ ■ ■ ■ ■ ■ ■

50

---

27. Because the *Mdn* is the score above and below which 50% of the raw scores lie, the interval which contains the *Mdn* is the interval which contains the score falling at the _____ of the frequency distribution.

■ ■ ■ ■ ■ ■ ■ ■ ■ ■ ■ ■ ■ ■

center

---

28. PLATE 6, page 13. The interval which contains the *Mdn* is ___ - ___ because the midpoint of the frequency distribution is contained in it.

■ ■ ■ ■ ■ ■ ■ ■ ■ ■ ■ ■ ■ ■

19-21

---

29. PLATE 6. The $\ell\ell$ of the interval containing the *Mdn* is _____ .

■ ■ ■ ■ ■ ■ ■ ■ ■ ■ ■ ■ ■ ■ ■

18.5

---

30. PLATE 6. 18.5 is the $\ell\ell$ to be used in Formula 1, because it is the lower limit of the interval ____ - ____ , which is the interval containing the *Mdn*.

■ ■ ■ ■ ■ ■ ■ ■ ■ ■ ■ ■ ■ ■ ■

19-21

---

31. PLATE 6. Σf_b = ____ because it is the sum of the frequencies in the intervals below the interval which contains the *Mdn*.

■ ■ ■ ■ ■ ■ ■ ■ ■ ■ ■ ■ ■ ■ ■

5

---

32. PLATE 6. Σf_b = 5 because it is the _____ of the f's which appear in the three class intervals below the interval which contains the *Mdn*.

■ ■ ■ ■ ■ ■ ■ ■ ■ ■ ■ ■ ■ ■ ■

sum

---

33. PLATE 6. f_w = ____ because it is the frequency within the interval 19-21 which contains the *Mdn*.

■ ■ ■ ■ ■ ■ ■ ■ ■ ■ ■ ■ ■ ■ ■

6

---

34. PLATE 6. In order to use Formula 1 to determine the *Mdn*, first determine the numerical values of: $\ell\ell$ = _____ , N = ____ , and Σf_b = ___ .

■ ■ ■ ■ ■ ■ ■ ■ ■ ■ ■ ■ ■ ■ ■

18.5, 19, 5

---

35. **PLATE 6.** $ll = 18.5$, $N = 19$, $\Sigma f_b = 5$. Also determine the numerical values of $f_w =$ ___, and $i =$ ___ .

■ ■ ■ ■ ■ ■ ■ ■ ■ ■ ■ ■ ■ ■ ■

6, 3

---

36. **PLATE 6.** $ll = 18.5$, $N = 19$, $\Sigma f_b = 5$, $f_w = 6$, $i = 3$. Substitute these numerical values for the symbols in Formula 1.

$$Mdn = \underline{\qquad} + \left(\frac{.5\underline{\qquad} - \underline{\quad}}{\underline{\qquad}} \right) \underline{\quad}$$

■ ■ ■ ■ ■ ■ ■ ■ ■ ■ ■ ■ ■ ■

Plate 10.

$$Mdn = 18.5 + \left(\frac{.5(19) - 5}{6} \right) 3$$

---

37. **PLATE 10.** Do the necessary arithmetic to determine the value of the *Mdn*.

$$Mdn = 18.5 + \left(\frac{.5(19) - 5}{6} \right) 3 = \underline{\qquad}$$

■ ■ ■ ■ ■ ■ ■ ■ ■ ■ ■ ■ ■ ■

20.75

---

38. **PLATE 6.** The *Mdn* is 20.75 because it is that point in the frequency distribution above which and below which ___ % of the raw scores lie.

■ ■ ■ ■ ■ ■ ■ ■ ■ ■ ■ ■ ■ ■

50

---

39. **PLATE 7, page 17.** The interval which contains the *Mdn* is ___ - ___ because the midpoint of the frequency distribution is contained in it.

■ ■ ■ ■ ■ ■ ■ ■ ■ ■ ■ ■ ■ ■

16-20

---

40. PLATE 7. The $\ell\ell$ of the interval containing the *Mdn* is _____ .

■ ■ ■ ■ ■ ■ ■ ■ ■ ■ ■ ■ ■ ■

15.5

---

41. PLATE 7. 15.5 is the $\ell\ell$ to be used in Formula 1 because it is the _____ (symbol) of the interval containing the *Mdn*.

■ ■ ■ ■ ■ ■ ■ ■ ■ ■ ■ ■ ■ ■

$\ell\ell$

---

42. PLATE 7. Σf_b = _____ because it is the sum of the f's below the interval 16-20.

■ ■ ■ ■ ■ ■ ■ ■ ■ ■ ■ ■ ■ ■

8

---

43. PLATE 7. f_w = _____ because it is the f of the interval 16-20 which contains the *Mdn*.

■ ■ ■ ■ ■ ■ ■ ■ ■ ■ ■ ■ ■ ■

6

---

44. PLATE 7. Σf_b = 8, f_w = 6. Also determine the numerical values: $\ell\ell$ = _____ , N = _____ , i = ___ .

■ ■ ■ ■ ■ ■ ■ ■ ■ ■ ■ ■ ■ ■

15.5, 21, 5

---

45. PLATE 7. $\ell\ell$ = 15.5, N = 21, Σf_b = 8, f_w = 6, i = 5. Substitute these numerical values for the symbols in Formula 1.

■ ■ ■ ■ ■ ■ ■ ■ ■ ■ ■ ■ ■ ■

Plate 11.

$$Mdn = 15.5 + \left(\frac{.5(21) - 8}{6} \right) 5$$

---

46. PLATE 11. Do the necessary arithmetic to determine the value of the *Mdn* of Plate 7. *Mdn* = _____.

■ ■ ■ ■ ■ ■ ■ ■ ■ ■ ■ ■ ■ ■ ■

17.58

---

47. PLATE 7. *Mdn* = 17.58 because it is that point in the frequency distribution above which and below which ____ % of the raw scores lie.

■ ■ ■ ■ ■ ■ ■ ■ ■ ■ ■ ■ ■ ■ ■

50

---

48. The computation of the medians of the distributions in Plates 6 and 7 have been for _____ (grouped/ungrouped) data.

■ ■ ■ ■ ■ ■ ■ ■ ■ ■ ■ ■ ■ ■ ■

grouped

---

49. Formula 1 can also be used for computing the *Mdn* for ungrouped data. Since, for ungrouped data, frequencies are presented for *each* score value, the *i* in Formula 1 for ungrouped data is equal to___.

■ ■ ■ ■ ■ ■ ■ ■ ■ ■ ■ ■ ■ ■ ■

1

---

50. PLATE 9. In order to compute the *Mdn*, determine:

$$ll = _____, N = \_\_\_\_\_, \Sigma f_b = \_\_\_.$$

■ ■ ■ ■ ■ ■ ■ ■ ■ ■ ■ ■ ■ ■ ■

87.5, 13, 6

---

51. PLATE 9. $ll = 87.5, N = 13, \Sigma f_b = 6$. Determine: $f_w = \_\_, i = \_\_$.

■ ■ ■ ■ ■ ■ ■ ■ ■ ■ ■ ■ ■ ■ ■

1, 1

52. PLATE 9. $\ell\ell = 87.5$, $N = 13$, $\Sigma f_b = 6$, $f_w = 1$, $i = 1$. Using Formula 1 determine the *Mdn* of the frequency distribution. *Mdn* = _____.

■ ■ ■ ■ ■ ■ ■ ■ ■ ■ ■ ■ ■ ■

88

Plate 12.

X	f
7	1
6	1
5	3
4	7
3	4
2	2
1	2

53. Determine the *Mdn* using Formula 1. *Mdn* = _____.

■ ■ ■ ■ ■ ■ ■ ■ ■ ■ ■ ■ ■ ■

3.79

Plate 13.

Interval	f
26-30	10
21-25	15
16-20	20
11-15	16
6-10	14
1-5	12

54. Determine the *Mdn* using Formula 1. *Mdn* = _____.

■ ■ ■ ■ ■ ■ ■ ■ ■ ■ ■ ■ ■ ■

15.88

55. The two measures of central tendency discussed thus far are the _____ and the _____.

■ ■ ■ ■ ■ ■ ■ ■ ■ ■ ■ ■ ■ ■

mode, median

---

56. The median differs from the mode in that the median is the _____ of the distribution, whereas the mode is the score value or class interval that has the largest _____.

■ ■ ■ ■ ■ ■ ■ ■ ■ ■ ■ ■ ■

midpoint, frequency

EXERCISES

1. What is meant by the term *mode* of a frequency distribution?

2. What is meant by the term *median* of a frequency distribution? What is its symbol?

3. Locate the *mode*, or modal interval, in the following frequency distributions. Locate the *median*, using Formula 1.

(a) X	f	(b) X	f	(c) Interval	f	(d) Interval	f
10	2	109	6	91-94	4	40-49	1
9	3	108	5	87-90	8	30-39	6
8	2	107	2	83-86	9	20-29	5
7	2	106	3	79-82	7	10-19	7
6	2	105	2	75-78	2	0-9	2
5	1	104	2				

4. What do the following symbols represent?

$$\Sigma \qquad \Sigma f_b \qquad f_w$$

MEASURES OF CENTRAL TENDENCY: MEAN

We shall now consider the third measure of central tendency—the mean. Unlike the mode and median, the mean lends itself to further and more sophisticated statistical analysis and is therefore the most useful of the three measures of central tendency. This set defines the mean and presents the formula for calculating it from both grouped and ungrouped data. This set will also describe a method for simplifying the computation of the mean by reducing the value of each of the raw scores.

SPECIFIC OBJECTIVES OF SET 3

At the conclusion of this set you will be able to:

(1) define the term *mean*.
(2) calculate the mean for ungrouped data, using Formula 2.
(3) determine the midpoint of a class interval.
(4) calculate the mean for grouped data, using Formula 2.
(5) calculate the mean for ungrouped data after reducing each raw score by a constant amount to simplify the calculation.
(6) calculate the mean for grouped data after reducing the midpoint of the class intervals by a constant amount to simplify the calculation.
(7) identify the symbol $\overline{X}$.

1. The third measure of central tendency, the *mean* is the arithmetic average of the raw scores. Thus, the average IQ score of a group of subjects is called the _____ of the group.

■ ■ ■ ■ ■ ■ ■ ■ ■ ■ ■ ■ ■ ■

mean

---

2. To obtain the mean, or arithmetic average, you sum all of the raw scores and divide by the _____ of raw scores involved.

■ ■ ■ ■ ■ ■ ■ ■ ■ ■ ■ ■ ■ ■

number

---

3. The symbol that represents the mean is $\overline{X}$. The ___ (symbol) represents the arithmetic average of the raw scores.

■ ■ ■ ■ ■ ■ ■ ■ ■ ■ ■ ■ ■ ■

$\overline{X}$

---

4. To obtain the mean ($\overline{X}$), first you must obtain the sum of all of the _____ _____ in the distribution.

■ ■ ■ ■ ■ ■ ■ ■ ■ ■ ■ ■ ■ ■

raw scores

---

5. To do this, you must multiply each value (X) by its frequency (f), which is written fX. Then you must sum these values. The symbol for this sum would be written ___ fX.

■ ■ ■ ■ ■ ■ ■ ■ ■ ■ ■ ■ ■ ■

Σ

---

6. To obtain the $\bar{X}$ you must divide the sum of the raw scores by the number of scores in the frequency distribution. Complete the formula for the mean by supplying the missing symbol.

$$\bar{X} = \frac{\Sigma fX}{\underline{\quad}}$$

■ ■ ■ ■ ■ ■ ■ ■ ■ ■ ■ ■ ■ ■

N

Formula 2. Calculation of the Mean

$$\bar{X} = \frac{\Sigma fX}{N}$$

7. Formula 2 presents the formula for the computation of the mean. The symbol ΣfX indicates that you multiply each score value by its f, and then _____ all of the products.

■ ■ ■ ■ ■ ■ ■ ■ ■ ■ ■ ■ ■ ■

sum

---

8. PLATE 4, page 10. There is an f of 3 for the score value 12. Therefore, in summing the raw scores, the value of 12 must be included _____ times.

■ ■ ■ ■ ■ ■ ■ ■ ■ ■ ■ ■ ■ ■

three

---

9. A simple method of determining the sum of the raw scores in a frequency distribution is to multiply each X by its f and then sum these products. This operation is represented in Formula 2 by the symbol ____.

■ ■ ■ ■ ■ ■ ■ ■ ■ ■ ■ ■ ■ ■

ΣfX

10. According to Formula 2, when you sum the raw scores and then divide by N, you obtain the _____ of the frequency distribution.

■ ■ ■ ■ ■ ■ ■ ■ ■ ■ ■ ■ ■ ■

mean

---

11. PLATE 8, page 21. To use Formula 2 in determining the $\bar{X}$ of the frequency distribution, you must first determine fX for each score value. To do this, you must multiply each score value by its __(symbol).

■ ■ ■ ■ ■ ■ ■ ■ ■ ■ ■ ■ ■ ■

f

---

12. PLATE 8. For the score value of 12, the $f = 3$; therefore, to determine fX for this score value, you must multiply _____ by__.

■ ■ ■ ■ ■ ■ ■ ■ ■ ■ ■ ■ ■ ■

12, 3

---

13. PLATE 8. The fX for the score value of 12 is ____ .

■ ■ ■ ■ ■ ■ ■ ■ ■ ■ ■ ■ ■ ■

36

---

14. PLATE 8. The fX for the score value of 12 is 36 because the f of this score value is __ .

■ ■ ■ ■ ■ ■ ■ ■ ■ ■ ■ ■ ■ ■

3

---

15. PLATE 8. For score value 11, $fX =$ _____ .

■ ■ ■ ■ ■ ■ ■ ■ ■ ■ ■ ■ ■ ■

zero

37

16. PLATE 8. The fX for score value 11 is zero because the f for this score value is _____ .

■ ■ ■ ■ ■ ■ ■ ■ ■ ■ ■ ■ ■ ■

zero

---

17. PLATE 8. For score value 10, fX = ____ .

■ ■ ■ ■ ■ ■ ■ ■ ■ ■ ■ ■ ■ ■

10

---

18. PLATE 8. The fX for score value 10 is 10 because the f of this score value is __ .

■ ■ ■ ■ ■ ■ ■ ■ ■ ■ ■ ■ ■ ■

1

---

19. PLATE 8. The values of fX for the score values of 9, 8, and 7 are__ , __ , and __ , respectively.

■ ■ ■ ■ ■ ■ ■ ■ ■ ■ ■ ■ ■ ■

9, 8, 7

---

20. PLATE 8. The fX for each of the score values of 9, 8, and 7 is 9, 8, and 7 because the f for each of these score values is __ .

■ ■ ■ ■ ■ ■ ■ ■ ■ ■ ■ ■ ■ ■

1

Plate 14.

X	f	fX
12	3	36
11	0	0
10	1	10
9	1	9
8	1	8
7	1	7
	$N = 7$	

38

21. This plate presents the frequency distribution of Plate 8 with an additional column giving the fX for each score value. Because Σ means "the sum of," the ΣfX for this distribution is the _____of all the fX values in the distribution.

■ ■ ■ ■ ■ ■ ■ ■ ■ ■ ■ ■ ■ ■

sum

22. PLATE 14. The ΣfX for the frequency distribution is ____ .

■ ■ ■ ■ ■ ■ ■ ■ ■ ■ ■ ■ ■ ■

70

23. PLATE 14. The ΣfX is 70 because it is the sum of each X times its __ (symbol).

■ ■ ■ ■ ■ ■ ■ ■ ■ ■ ■ ■ ■ ■

f

24. Formula 2 indicates that to obtain the mean ($\bar{X}$) it is necessary that ΣfX be divided by____ (symbol).

■ ■ ■ ■ ■ ■ ■ ■ ■ ■ ■ ■ ■ ■

N

25. PLATE 14. $\Sigma fX = 70$, $N = 7$. Using Formula 2, substitute these numerical values for the symbols and calculate the mean.

$$\bar{X} = \frac{\Sigma fX}{N} = \frac{\quad\quad}{\quad\quad} = \underline{\quad}$$

■ ■ ■ ■ ■ ■ ■ ■ ■ ■ ■ ■ ■ ■

$$\frac{70}{7} = 10$$

26. PLATE 12, page 31. To determine the $\bar{X}$ of the distribution, first determine fX for each score value.

■ ■ ■ ■ ■ ■ ■ ■ ■ ■ ■ ■ ■ ■

7, 6, 15, 28, 12, 4, 2

---

27. PLATE 12. The fX for each of the score values is 7, 6, 15, 28, 12, 4, and 2. The $\Sigma fX =$ _____ because it is the sum of the fX for each score value.

■ ■ ■ ■ ■ ■ ■ ■ ■ ■ ■ ■ ■ ■

74

---

28. PLATE 12. $N =$ _____ .

■ ■ ■ ■ ■ ■ ■ ■ ■ ■ ■ ■ ■ ■

20

---

29. PLATE 12. $\Sigma fX = 74$, $N = 20$. Using Formula 2, determine the $\bar{X}$ of the distribution.

$$\bar{X} = \underline{\qquad}$$

■ ■ ■ ■ ■ ■ ■ ■ ■ ■ ■ ■ ■ ■

3.7

Plate 15.

X	f	fX
25	2	___
24	1	___
23	3	___
22	1	___
21	4	___
	$N = 11$	

40

30. To determine the $\bar{X}$ of the distribution, first determine fX for each score value and enter these values above.

■ ■ ■ ■ ■ ■ ■ ■ ■ ■ ■ ■ ■ ■

50, 24, 69, 22, 84

---

31. PLATE 15. The ΣfX is _____ because it is the sum of the fX for each score value.

■ ■ ■ ■ ■ ■ ■ ■ ■ ■ ■ ■ ■ ■

249

---

32. PLATE 15. $N = 11$, $\Sigma fX = 249$. Use Formula 2 to determine the mean of the distribution. $\bar{X} =$ _____ .

■ ■ ■ ■ ■ ■ ■ ■ ■ ■ ■ ■ ■ ■

22.6

---

33. To determine the $\bar{X}$ more simply, a *constant* amount may be subtracted from each score value before computing ΣfX, then added to the obtained mean.

PLATE 15. If you subtract 20 from each of the five score values, you obtain the five score values of __, __, __, __, and __.

■ ■ ■ ■ ■ ■ ■ ■ ■ ■ ■ ■ ■ ■

5, 4, 3, 2, 1

Plate 16. Data of Plate 15 with Each X Reduced by Twenty Points

$(X - 20)$	f	$f(X - 20)$
5	2	
4	1	
3	3	
2	1	
1	4	
$N = 11$		$\Sigma f(X - 20) =$ ___

41

34. This plate presents the data of Plate 15 with a *constant* of 20 subtracted from each score value. Compute $f(X - 20)$ for each score value in Plate 16 and then determine the $\Sigma f(X - 20)$ for this distribution.

■ ■ ■ ■ ■ ■ ■ ■ ■ ■ ■ ■ ■ ■

10, 4, 9, 2, 4; $\Sigma f(X - 20) = 29$

---

35. PLATE 16. $N = 11$, $\Sigma f(X - 20) = 29$.

To determine the $\bar{X}$ of a frequency distribution in which a *constant* amount has been subtracted from each score value, it is necessary to add the constant to the obtained mean. The obtained mean in Plate 16 is _____ .

■ ■ ■ ■ ■ ■ ■ ■ ■ ■ ■ ■ ■ ■

2.6

---

36. PLATE 16. Remember, you have subtracted a constant of 20 points from each score value, so it is necessary to add 20 to the obtained mean in order to determine the true mean of the distribution.

$$\bar{X} = \underline{\quad} + \underline{\quad} = \underline{\quad}$$

■ ■ ■ ■ ■ ■ ■ ■ ■ ■ ■ ■ ■ ■

2.6 + 20 = 22.6

---

37. To compute the $\bar{X}$ of a frequency distribution containing grouped data, you must first determine the one score value that best represents each interval. The midpoint of each interval is used for this purpose. For interval 16-20, the midpoint is _____ .

■ ■ ■ ■ ■ ■ ■ ■ ■ ■ ■ ■ ■

18

---

38. The midpoint of interval 16-20 is 18; and therefore, 18 is used as the score value that best _____ the interval.

■ ■ ■ ■ ■ ■ ■ ■ ■ ■ ■ ■ ■

represents

42

39. The midpoint of interval 1-5 is ___ .

■ ■ ■ ■ ■ ■ ■ ■ ■ ■ ■ ■ ■ ■

3

---

40. The midpoint of interval 1-5 is 3 because _____ score values lie above it and _____ score values lie below it.

1 2 3 4 5
↑
midpoint

■ ■ ■ ■ ■ ■ ■ ■ ■ ■ ■ ■ ■ ■

two, two

---

41. The midpoint of interval 1-6 is 3.5 because _____ score values lie above it and _____ lie below it.

1 2 3 4 5 6
↑
midpoint

■ ■ ■ ■ ■ ■ ■ ■ ■ ■ ■ ■ ■ ■

three, three

---

42. When the interval encompasses an *even* number of score values, the midpoint always ends in .5. When the interval encompasses an *odd* number of score values, the midpoint is always the _____ score value of the interval.

■ ■ ■ ■ ■ ■ ■ ■ ■ ■ ■ ■ ■ ■

middle

---

43. PLATE 6, page 13. To determine the $\bar{X}$ of the frequency distribution, first determine the midpoint of each interval in the distribution. The midpoints are ___ , ___ , ___ , ___ , ___ , ___ , and ___ .

■ ■ ■ ■ ■ ■ ■ ■ ■ ■ ■ ■ ■ ■

29, 26, 23, 20, 17, 14, 11

---

44. PLATE 6, page 13. The midpoint of each interval is used as the X value in Formula 2 because these midpoints are the score values that best _____ the score values encompassed by the intervals.

■ ■ ■ ■ ■ ■ ■ ■ ■ ■ ■ ■ ■ ■

represent

Plate 17.

Interval	X	f
28–30	29	2
25–27	26	4
22–24	23	2
19–21	20	6
16–18	17	2
13–15	14	1
10–12	11	2
		$N = 19$

45. This plate presents the frequency distribution of Plate 6 and an additional column, X, which is the _____ of each interval.

■ ■ ■ ■ ■ ■ ■ ■ ■ ■ ■ ■ ■

midpoint

---

46. For grouped data, the midpoint of each interval is used as ___ (symbol) in Formula 2 to determine the mean.

■ ■ ■ ■ ■ ■ ■ ■ ■ ■ ■ ■ ■

X

---

47. PLATE 17. (1) Determine each fX.
    (2) $\Sigma fX =$ ____
    (3) Using Formula 2, $\bar{X} =$ ____

■ ■ ■ ■ ■ ■ ■ ■ ■ ■ ■ ■ ■

(1) 58, 104, 46, 120, 34, 14, 22
(2) 398
(3) 20.9

48. The method presented earlier for simplifying the arithmetic involved in the calculation of the $\bar{X}$ was to reduce each score value by a constant amount. This can also be done with grouped data by reducing the _____ of each interval by a constant amount.

■ ■ ■ ■ ■ ■ ■ ■ ■ ■ ■ ■ ■ ■ ■

midpoint

49. PLATE 17. Each X value could be reduced by a constant of 10, in which case the obtained mean would have to be _____ (increased/decreased) by 10.

■ ■ ■ ■ ■ ■ ■ ■ ■ ■ ■ ■ ■ ■ ■

increased

Plate 18.

interval	X	$(X-10)$	f	$f(X-10)$
28–30	29	19	2	38
25–27	26	16	4	64
22–24	23	13	2	26
19–21	20	10	6	60
16–18	17	7	2	14
13–15	14	4	1	4
10–12	11	1	2	2
			$N = 19$	$\Sigma f(X-10) = 208$

$$\text{obtained } \bar{X} = \frac{208}{19} = 10.9$$

$$\text{true } \bar{X} = 10.9 + 10 = 20.9$$

50. This plate presents the same data as Plate 17, but each X value has been reduced by 10. The reduced X values are presented in the third column of the plate. The obtained mean has been increased by ____ in order to obtain the true mean.

■ ■ ■ ■ ■ ■ ■ ■ ■ ■ ■ ■ ■ ■

10

51. PLATE 18. The $\bar{X}$ =_____, which is the same as that obtained by the longer method when computing from the distribution presented in Plate 17.

■ ■ ■ ■ ■ ■ ■ ■ ■ ■ ■ ■ ■

20.9

Plate 19. Raw Scores

20	28	25
26	30	26
21	34	31
29	22	28
32	24	

52. Using Plate 18 as a guide, prepare a frequency distribution of the raw scores presented in Plate 19. Use an interval of three score values each. Simplify your calculations by using a constant of 20. Compute the $\bar{X}$ of the distribution.

interval	X	(X − 20)	f	f (X − 20)

■ ■ ■ ■ ■ ■ ■ ■ ■ ■ ■ ■ ■

Plate 20.

interval	X	$(X-20)$	f	$f(X-20)$
32–34	33	13	2	26
29–31	30	10	3	30
26–28	27	7	4	28
23–25	24	4	2	8
20–22	21	1	3	3
			$N = 14$	$\Sigma f(X-20) = 95$

$$\overline{X} = \frac{95}{14} + 20 = 6.79 + 20 = 26.79$$

1. What is meant by the term *mean* of a frequency distribution. What symbol denotes *mean*?

2. Determine the midpoint of each of the following intervals.

    1–5   5–6   101–110   80–84

3. Calculate the mean for the following frequency distributions, using Formula 2.

(a) X	f	(b) X	f	(c) interval	f	(d) interval	f
9	1	59	3	21–25	4	57–60	2
8	2	58	4	16–20	9	53–56	5
7	4	57	6	11–15	10	49–52	9
6	3	56	2	6–10	3	45–48	3
5	2	55	2	1–5	2	41–44	2
4	2						

4. Simplify the calculation of the mean of the frequency distribution presented in 3(b) by subtracting a constant of 50 from each raw score. Check your answer with your answer to 3(b).

5. Simplify the calculation of the mean of the frequency distribution presented in 3(d) by subtracting a constant of 40 from each class interval. Check your answer with your answer to 3(d).

COMPARISON OF MEASURES OF CENTRAL TENDENCY/ PERCENTILES

We have just shown how we can simplify our calculations by subtracting a constant from each raw score. We can make similar adjustments by adding, multiplying, and dividing by a constant, as will be shown in the first part of this set. The procedure of using constants can eliminate computation with negative scores.

The mode, median, and mean are affected differently by changes in extreme scores in the frequency distribution. This set will also compare the ways in which these three measures of central tendency are affected by the alteration of extreme scores.

It is sometimes useful to convert a raw score into a percentile rank which tells what percentage of scores lies below it. The procedure for obtaining the percentile rank of a raw score in a distribution is given, as well as the definition of quartiles and deciles.

SPECIFIC OBJECTIVES OF SET 4

At the conclusion of this set you will be able to:

(1) simplify the calculation of the mean of a frequency distribution containing negative scores by adding a constant to each score.

(2) determine the effect of extreme score values on the value of the mean, median, and mode.

(3) define the term *percentile*.

(4) use Formula 1 to determine the score which lies at a given percentile for grouped data.

(5) use Formula 1 to determine the score which lies at a given percentile for ungrouped data.

(6) define the terms *quartile* and *decile*.

(7) use Formula 3 to determine the percentile of a specific score in a frequency distribution.

(8) identify the symbols Q_1, Q_2, Q_3, D_1.

1. Thus far you have been *subtracting* a constant from each score value in order to simplify the calculation of $\overline{X}$. You may also add a constant. Of course, when you add a constant, you must _____ the same constant from the obtained mean to determine the true mean.

■ ■ ■ ■ ■ ■ ■ ■ ■ ■ ■ ■ ■ ■

subtract

2. You may also *multiply* each score value by a constant. In this case, you must _____ the obtained mean by the constant in order to obtain the true mean.

■ ■ ■ ■ ■ ■ ■ ■ ■ ■ ■ ■ ■ ■

divide

3. If you *divide* each score value by a constant, you must _____ the obtained mean by the constant in order to obtain the true mean.

■ ■ ■ ■ ■ ■ ■ ■ ■ ■ ■ ■ ■ ■

multiply

4. You have used only positive numbers thus far. However, you may do all statistical calculations with negative numbers as well as positive ones as long as you handle the negative numbers properly. Thus: the ℓℓ of score value −19 is −19.5; the ℓℓ of score value −4 is _____ .

■ ■ ■ ■ ■ ■ ■ ■ ■ ■ ■ ■ ■ ■

−4.5

5. In cases where negative numbers make the computation difficult, it is desirable to _____ a constant to the score value in order to make them all positive.

■ ■ ■ ■ ■ ■ ■ ■ ■ ■ ■ ■ ■ ■

add

51

6. If the score value is -7 and you add a constant of 10, the score value becomes___.

■ ■ ■ ■ ■ ■ ■ ■ ■ ■ ■ ■ ■

3

Plate 21. Raw Scores

1	2
0	2
-1	1
4	-1
-2	3
	1

7. Prepare a frequency distribution, adding a constant of 3 to each score value, and compute the $\bar{X}$.

X	$(X + 3)$	f	$f(X + 3)$

■ ■ ■ ■ ■ ■ ■ ■ ■ ■ ■ ■ ■

Plate 22.

X	$(X + 3)$	f	$f(X + 3)$
4	7	1	7
3	6	1	6
2	5	2	10
1	4	3	12
0	3	1	3
−1	2	2	4
−2	1	1	1

$$N = 11 \quad \Sigma f(X+3) = 43$$

obtained $\bar{X} = 3.9$

true $\bar{X} = 3.9 - 3 = .9$

---

8. PLATE 22. The constant of 3 was subtracted from the obtained $\bar{X}$ because it was _____ as a constant to each of the score values.

■ ■ ■ ■ ■ ■ ■ ■ ■ ■ ■ ■ ■

added

---

9. PLATE 22. The mode of the distribution of X values is __.

■ ■ ■ ■ ■ ■ ■ ■ ■ ■ ■ ■ ■

1

---

10. PLATE 22. 1 is the mode because it is the most _____ score value in the distribution.

■ ■ ■ ■ ■ ■ ■ ■ ■ ■ ■ ■ ■

frequent

---

11. PLATE 22. Determine the *Mdn* of the distribution, using Formula 1. (Do not use the $(X+3)$ column. Instead use the original X values.)

Mdn = _____

■ ■ ■ ■ ■ ■ ■ ■ ■ ■ ■ ■ ■

Plate 23.

$$Mdn = .5 + \left(\frac{.5(11) - 4}{3}\right)1 = 1$$

---

12. The *Mdn* is the score value above which and below which ____ % of the raw scores lie.

■ ■ ■ ■ ■ ■ ■ ■ ■ ■ ■ ■

50

Plate 24.

(a)			(b)	
X	*f*		*X*	*f*
7	1		12	1
6	2		11	0
5	2		10	0
4	3		9	0
3	4		8	0
2	2		7	0
1	1		6	2
	N = 15		5	2
			4	3
			3	4
			2	2
			1	1
				N = 15

13. Compare the two distributions presented in Plate 24. The top score of 7 in (24a) has been changed to ____ in (24b). All the other score values and their *f*'s have remained the same.

■ ■ ■ ■ ■ ■ ■ ■ ■ ■ ■ ■

12

54

14. PLATE 24. The mode for (24a) is ___. This is _____ (the same as/different from) the mode for (24b).

■ ■ ■ ■ ■ ■ ■ ■ ■ ■ ■ ■ ■ ■

3, the same as

15. PLATE 24. The *Mdn* of (24a) and (24b) is _____ (the same/different).

■ ■ ■ ■ ■ ■ ■ ■ ■ ■ ■ ■ ■ ■

the same

16. PLATE 24. Of the three measures of central tendency, the only one affected by the change in the raw score of 7 in (24a) to 12 in (24b) is the ___ (symbol).

■ ■ ■ ■ ■ ■ ■ ■ ■ ■ ■ ■ ■ ■

$\bar{X}$

17. Consider the raw scores: 7 8 9 9 9 10 11. The mode of this set of raw scores is ___. The *Mdn* is ___ . The $\bar{X}$ is ___.

■ ■ ■ ■ ■ ■ ■ ■ ■ ■ ■ ■ ■ ■

9, 9, 9

18. Now if the score of 11 becomes 20; that is, if 7 8 9 9 9 10 11 becomes 7 8 9 9 9 10 20, the mode _____ (is the same/becomes larger/becomes smaller), the *Mdn* _____ (is the same/becomes larger/becomes smaller), the $\bar{X}$ _____ (is the same/ becomes larger/becomes smaller).

■ ■ ■ ■ ■ ■ ■ ■ ■ ■ ■ ■ ■

is the same, is the same, becomes larger

19. Because the *Mdn* is defined as the point above which and below which 50% of the scores lie, it _____ (is/is not) influenced by the numerical value of the scores above or below it.

■ ■ ■ ■ ■ ■ ■ ■ ■ ■ ■ ■ ■ ■

is not

20. If you add 10 points to a score above the *Mdn*, it _____ (will/will not) change the value of the *Mdn*.

■ ■ ■ ■ ■ ■ ■ ■ ■ ■ ■ ■ ■ ■

will not

21. Since the $\bar{X}$ is determined by summing the raw scores and dividing by *N*, if you add 10 points to a score above the $\bar{X}$, the value of the $\bar{X}$ will _____ _____ (be decreased/be increased/remain the same).

■ ■ ■ ■ ■ ■ ■ ■ ■ ■ ■ ■ ■ ■

be increased

22. The _____ is affected by the value of each of the raw scores, whereas the _____ is concerned only with the ordering of the raw scores. The _____ is concerned only with which score occurs most frequently.

■ ■ ■ ■ ■ ■ ■ ■ ■ ■ ■ ■ ■ ■

mean, median, mode

23. You may desire to describe a point in the frequency distribution in terms of the percentage of scores falling below it. The term *percentile* is used to describe the score below which a given _____ of the total number of scores lie.

■ ■ ■ ■ ■ ■ ■ ■ ■ ■ ■ ■ ■ ■

percentage

24. Thus, the 20th percentile is the score below which 20% of the scores lie. 40% of the scores lie below the score that is at the _____ _____ .

■ ■ ■ ■ ■ ■ ■ ■ ■ ■ ■ ■ ■ ■ ■

40th percentile

25. Because the *Mdn* is the score value below which 50% of the scores lie, it is also termed the _____ _____ of the distribution.

■ ■ ■ ■ ■ ■ ■ ■ ■ ■ ■ ■ ■ ■ ■

50th percentile

Plate 25.

Interval	f
26-30	3
21-25	3
16-20	5
11-15	4
6-10	3
1-5	2
	$N = 20$

26. If you wish to determine the score value that represents the 40th percentile, you must first determine how many scores are 40% of the total number of scores. In this case, $N = 20$. 40% of 20 is___ .

■ ■ ■ ■ ■ ■ ■ ■ ■ ■ ■ ■ ■ ■ ■

8

27. PLATE 25. Eight scores represent 40% of the total number of scores. Starting from the bottom of the distribution, sum the f column until you determine which interval contains the eighth score from the bottom. Interval ____-____ contains the eighth score from the bottom of the distribution.

■ ■ ■ ■ ■ ■ ■ ■ ■ ■ ■ ■ ■ ■ ■

11-15

28. **PLATE 25.** The eighth score, which is at the 40th percentile, is contained in interval 11-15. To determine the exact score value of the 40th percentile, you may use Formula 1. However, you are now seeking the 40th percentile instead of the 50th percentile (*Mdn*) so you must change .5N to .___N.

■ ■ ■ ■ ■ ■ ■ ■ ■ ■ ■ ■ ■ ■

.4

29. Substituting the value .4N for the 40th percentile, and the term "40th percentile" for the symbol "*Mdn*," Formula 1 reads:

$$\text{40th percentile} = \ell\ell + \left(\frac{.4N - \Sigma f_b}{f_w} \right) i$$

PLATE 25. Determine the 40th percentile. Recall that it is located in interval 11-15.

$$\text{40th percentile} = \underline{\hspace{2cm}}$$

■ ■ ■ ■ ■ ■ ■ ■ ■ ■ ■ ■ ■ ■

Plate 26.

$$\text{40th percentile} = 10.5 + \left(\frac{.4(20) - 5}{4} \right) 5 = 14.25$$

30. Write the formula for determining the 25th percentile of a frequency distribution. Making the necessary changes, use Formula 1 as a guide.

$$\text{25th percentile} = \underline{\hspace{4cm}}$$

■ ■ ■ ■ ■ ■ ■ ■ ■ ■ ■ ■ ■ ■

$$\text{25th percentile} = \ell\ell + \left(\frac{.25N - \Sigma f_b}{f_w} \right) i$$

31. Formula 1 can be adapted for use in determining any percentile as long as you use the decimal form of the percentage when multiplying by the ___ (symbol).

■ ■ ■ ■ ■ ■ ■ ■ ■ ■ ■ ■ ■ ■

N

---

32. In determining a percentile for Plate 25, you were using a frequency distribution for _____ (grouped/ungrouped) data.

■ ■ ■ ■ ■ ■ ■ ■ ■ ■ ■ ■ ■ ■

grouped

---

33. When determining percentiles for a frequency distribution of ungrouped data, the i in Formula 1 is equal to ___ .

■ ■ ■ ■ ■ ■ ■ ■ ■ ■ ■ ■ ■ ■

1

---

34. PLATE 3. Determine the 30th percentile of the distribution. (Substitute the correct numerical values for this calculation in Formula 1.)

30th percentile = _____

■ ■ ■ ■ ■ ■ ■ ■ ■ ■ ■ ■ ■

Plate 27.

$$30\text{th percentile} = 100.5 + \left(\frac{.3(16) - 3}{3} \right) 1 = 101.1$$

---

35. The most commonly used percentiles are the 25th percentile, the 50th percentile, and the 75th percentile. The score at the 25th percentile is known as the *1st quartile* (symbol Q_1) because one quarter of the scores lie _____ it.

■ ■ ■ ■ ■ ■ ■ ■ ■ ■ ■ ■ ■ ■

below

---

36. The 2nd quartile (symbol Q_2) is the point below which 50% of the scores lie, and is also called the _____ (symbol).

■ ■ ■ ■ ■ ■ ■ ■ ■ ■ ■ ■ ■ ■

Mdn

37. The 3rd quartile (symbol Q_3) is the score at the _____ percentile, and _____ % of the scores lie below it.

■ ■ ■ ■ ■ ■ ■ ■ ■ ■ ■ ■ ■ ■

75th, 75

38. The symbol for the 1st quartile is Q_1. The symbols for the 2nd and 3rd quartiles are _____ and _____ .

■ ■ ■ ■ ■ ■ ■ ■ ■ ■ ■ ■ ■ ■

Q_2, Q_3

39. 25% of the scores lie between Q_1 and Q_2. _____ % of the scores lie between Q_2 and Q_3. _____ % of the scores lie between Q_1 and Q_3.

■ ■ ■ ■ ■ ■ ■ ■ ■ ■ ■ ■ ■ ■

25, 50

40. The median lies at the _____ percentile, which is also _____ (use symbol for the correct quartile).

■ ■ ■ ■ ■ ■ ■ ■ ■ ■ ■ ■ ■ ■

50th, Q_2

41. In addition to *quartiles*, percentiles are sometimes referred to in terms of deciles. Since the word "decimal" means 10, the *1st decile* must lie at the _____ percentile.

■ ■ ■ ■ ■ ■ ■ ■ ■ ■ ■ ■ ■ ■

10th

42. The symbol used for decile is D. D_1 represents the lst decile, or 10th percentile. D_2 represents the 2nd decile, or 20th percentile. The symbol for the 4th decile is _____ (symbol) and it lies at the _____ percentile.

■ ■ ■ ■ ■ ■ ■ ■ ■ ■ ■ ■ ■ ■

D_4, 40th

43. The 50th percentile lies at the same score value as _____ (give symbol for the correct decile). It also lies at the same score value as _____ (give symbol for the correct quartile) and is known as the _____ (give symbol for the correct measure of central tendency).

■ ■ ■ ■ ■ ■ ■ ■ ■ ■ ■ ■ ■ ■

D_5, Q_2, *Mdn*

Formula 3. Calculation of the Percentile

$$\text{Percentile in decimal form} \quad \frac{\left(\dfrac{X - \ell\ell}{i}\right) f_w + \Sigma f_b}{N}$$

in which: X = score value for which the percentile is to be computed
f_b = frequency below the i which contains the score
f_w = frequency within the i which contains the score

Multiply the decimal form of the percentile by 100 in order to determine the percentile.

---

44. Formula 1 permits you to determine the score value that lies at a particular percentile. When you wish to determine the percentile equivalent for a particular raw score, you use Formula 3. In Formula 3 the score value for which the percentile is to be computed is represented by____(symbol).

■ ■ ■ ■ ■ ■ ■ ■ ■ ■ ■ ■ ■ ■

X

---

45. To use Formula 3 in determining the percentile equivalent of a particular score, first you must determine the interval within which the score lies.

PLATE 25. The raw score of 22 lies in i ____-____.

■ ■ ■ ■ ■ ■ ■ ■ ■ ■ ■ ■ ■ ■

21-25

---

46. PLATE 25. To determine the percentile equivalent of raw score 22, which lies in i 21-25, substitute the numerical values for the symbols in Formula 3.

$$\text{Percentile in decimal form} = \frac{\left(\dfrac{\underline{\quad} - \underline{\quad}}{\underline{\quad}}\right)\underline{\quad} + \underline{\quad}}{\underline{\quad}}$$

■ ■ ■ ■ ■ ■ ■ ■ ■ ■ ■ ■ ■ ■

Plate 28.

$$\text{Percentile in decimal form} = \frac{\left(\dfrac{22 - 20.5}{5}\right)3 + 14}{20}$$

---

47. PLATE 28. The percentile equivalent in decimal form of raw score 22 is _____ (to three decimal places).

■ ■ ■ ■ ■ ■ ■ ■ ■ ■ ■ ■ ■ ■

.745

---

48. PLATE 25. .745 is the decimal form of the percentile equivalent of raw score 22. To convert the decimal form into a percentage, you must multiply the decimal form by _____. Raw score 22 lies at the _____ percentile of the distribution.

■ ■ ■ ■ ■ ■ ■ ■ ■ ■ ■ ■ ■ ■

100, 74.5th

---

49. PLATE 6, page 13. The percentile equivalent of raw score 20 is_____. (Substitute, in Formula 3, the correct numerical values for this calculation.)

■ ■ ■ ■ ■ ■ ■ ■ ■ ■ ■ ■ ■ ■

Plate 29.

$$\text{Percentile in decimal form} = \frac{\left(\dfrac{20 - 18.5}{3}\right)6 + 5}{19} = .42$$

$$\text{Percentile} = .42(100) = 42\text{nd percentile}$$

50. The percentile equivalent of any particular raw score indicates the percentage of scores falling _____ this particular raw score.

■ ■ ■ ■ ■ ■ ■ ■ ■ ■ ■ ■ ■

below

---

51. The percentile at which a particular raw score lies will tell you the _____ of scores lying below it.

■ ■ ■ ■ ■ ■ ■ ■ ■ ■ ■ ■ ■

percentage

---

52. A raw score of 97 has no particular meaning because it tells you nothing about its position relative to the other raw scores in the distribution. However, to say that it is at the 75th percentile, or at Q_3, tells you that below it lie ____ % of the raw scores and above it lie ____ % of the raw scores.

■ ■ ■ ■ ■ ■ ■ ■ ■ ■ ■ ■ ■

75, 25

EXERCISES

1. Calculate the $\bar{X}$ of the following distribution. Simplify your calculations by adding a constant of 5.

X	f
2	2
1	3
0	5
−1	4
−2	3
−3	3
−4	1

2. Which measure or measures of central tendency are affected by altering the values of extreme scores in a frequency distribution? Which are not affected?

3. If John Jones has a language achievement score at the 80th percentile, how can you interpret his score in respect to the total group?

4. What percentile lies at the first quartile? The second quartile? The third quartile? What symbols denote these quartiles?

5. What percentile lies at the first decile? The fourth decile? The sixth decile? What symbols denote these deciles?

6. For the following distributions, determine the score values which lie at the 40th percentile, at Q_3, at D_6. Use the modification of Formula 1 for your calculations.

(a) X	f	(b) interval	f
15	1	17-20	4
14	2	13-16	6
13	4	9-12	9
12	6	5-8	4
11	3	1-4	2
10	3		
9	1		

7. Determine the percentiles of score values 11, 13, and 14 in Exercise 6(a), using Formula 3.

8. Determine the percentiles of score values 6, 12, and 17 in Exercise 6(b), using Formula 3.

GRAPHIC REPRESENTATION
OF FREQUENCY DISTRIBUTIONS

The score value which represents the central tendency of a frequency distribution tells us nothing about the spread of the scores around that point. Even when a frequency distribution of scores is prepared it is difficult to see the "shape" of the distribution. To determine the manner in which the scores are spread it is often helpful to present them in graphic form. In this set you will be shown how to prepare *bar histograms* of frequency distributions for both grouped and ungrouped data. Here you will be introduced to the concept of a "smoothed" curve and given definitions of bimodal and unimodal distributions, and symmetrical, negatively skewed, and positively skewed distributions. You will learn how the mean, median, and mode are related to the shape of the frequency polygon and how they are affected by the skewness of the distribution.

SPECIFIC OBJECTIVES OF SET 5

At the conclusion of this set you will be able to:

(1) draw bar histograms of grouped and ungrouped data.

(2) draw a frequency polygon of grouped and ungrouped data.

(3) identify bimodal, unimodal, symmetrical, and negatively and positively skewed frequency polygons.

(4) describe the relationship of mean, median, and mode to the shape of a frequency polygon.

1. In order to present an over-all picture of a frequency distribution, it is often desirable to depict it in graphic form. Plate 30 presents a frequency distribution for grouped data.

Plate 30.

interval	f	real limits of interval
13-15	4	_____._____
10-12	8	_____._____
7-9	9	_____._____
4-6	7	_____._____
1-3	2	_____._____

In Set 1 we learned that the real limits of an interval lie .5 above and .5 below the interval. We need the real limits of each interval in order to prepare a histogram. Enter these real limits in Plate 30.

■ ■ ■ ■ ■ ■ ■ ■ ■ ■ ■ ■ ■ ■

12.5-15.5, 9.5-12.5, 6.5-9.5, 3.5-6.5, .5-3.5

The histogram for the distribution in Plate 30 is presented below.

Plate 31.

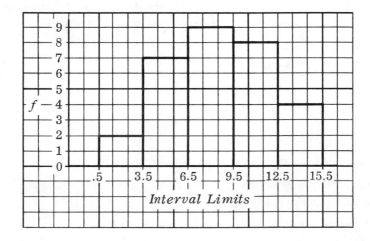

66

2. Notice that the real limits or the intervals are designated along the _____ (vertical/horizontal) axis and the frequency of scores within the intervals are designated along the _____ axis.

■ ■ ■ ■ ■ ■ ■ ■ ■ ■ ■ ■ ■ ■

horizontal, vertical

3. Each interval is represented by a vertical bar. For this reason the graph is often referred to as a *bar histogram*. The *f* for any given interval is shown by the _____ of the bar.

■ ■ ■ ■ ■ ■ ■ ■ ■ ■ ■ ■ ■ ■

height

4. As illustrated in Plate 31, in constructing a histogram, the lowest value in the distribution usually is presented on the _____ (left/right) of the horizontal axis.

■ ■ ■ ■ ■ ■ ■ ■ ■ ■ ■ ■ ■ ■

left

5. To get an accurate graphic representation of the distribution, the frequency scale along the vertical axis must show the _____ point at the bottom of the scale, as in Plate 31.

■ ■ ■ ■ ■ ■ ■ ■ ■ ■ ■ ■ ■ ■

zero

6. Construct the histogram for the following set of data:

Plate 32.

interval	f
41-45	1
36-40	3
31-35	5
26-30	6
21-25	7
16-20	4
11-15	2

■ ■ ■ ■ ■ ■ ■ ■ ■ ■ ■ ■ ■ ■

Plate 33.

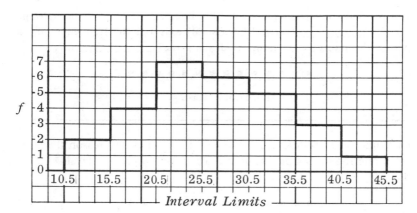

7. The method above for graphically depicting a frequency distribution is satisfactory when dealing with only one distribution. However, if we wished to compare a number of distributions in the same graph, the overlapping bars would be confusing. Another graphic method, the *frequency polygon*, will help in such cases. It utilizes the midpoints of the intervals. In Plate 32, these midpoints are 43, 38, 33, 28, ____ , ____ , and ____ .

■ ■ ■ ■ ■ ■ ■ ■ ■ ■ ■ ■ ■ ■

23, 18, 13

8. If we draw a line connecting the midpoints of the tops of each bar in the histogram, we will, in effect, be constructing a frequency polygon. Draw this line in Plate 33.

■ ■ ■ ■ ■ ■ ■ ■ ■ ■ ■ ■ ■ ■

Plate 34.

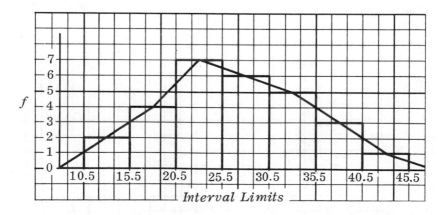

9. Notice in Plate 34 that, for intervals having a frequency of zero, the point representing that frequency will lie on the _____ axis. There is always such a point at each _____ of the distribution.

■ ■ ■ ■ ■ ■ ■ ■ ■ ■ ■ ■ ■ ■

horizontal, end

---

10. Thus, in a frequency polygon, a point is located directly above the _____ of the interval, such that its vertical distance from the horizontal axis represents the _____ in that interval.

■ ■ ■ ■ ■ ■ ■ ■ ■ ■ ■ ■ ■ ■

midpoint, frequency

---

11. For ungrouped data, each value should be considered as an interval with its own upper and lower limits. For example, for value 5, the real limits are 4.5 and _____. In constructing a frequency polygon, the midpoint of this interval is, of course, ___.

■ ■ ■ ■ ■ ■ ■ ■ ■ ■ ■ ■ ■

5.5, 5

---

12. Therefore, for ungrouped data, we merely indicate the values along the horizontal axis, keeping in mind that each value is, in reality, an _____ having upper and lower real limits.

■ ■ ■ ■ ■ ■ ■ ■ ■ ■ ■ ■ ■

interval

Plate 35. Frequency Distribution

X	f
90	1
89	3
88	2
87	3
86	4
85	2
84	1
	$N = 16$

13. Construct the frequency polygon for the data in Plate 35. (Do not construct the histogram.)

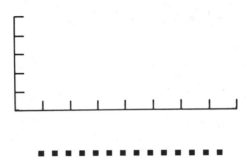

■ ■ ■ ■ ■ ■ ■ ■ ■ ■ ■ ■ ■ ■

Plate 36.

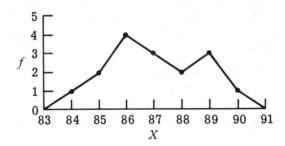

14. The graphic representation of the frequency distribution in Plate 36 is called a _____ _____ .

frequency polygon

15. In review, along the horizontal axis, the score values are presented with the lowest value at the _____ (left/right) and the highest value at the _____ (left/right).

■ ■ ■ ■ ■ ■ ■ ■ ■ ■ ■ ■ ■ ■

left, right

---

16. Along the vertical axis, the f's of the score values are presented with the lowest f at the _____ (bottom/top) of the axis and the highest f at the _____ (bottom/top) of the axis.

■ ■ ■ ■ ■ ■ ■ ■ ■ ■ ■ ■ ■ ■

bottom, top

---

17. Thus, the lowest values of both the horizontal and vertical axis in a frequency polygon are located in the _____ _____ corner of the graph.

■ ■ ■ ■ ■ ■ ■ ■ ■ ■ ■ ■ ■ ■

lower left

---

18. PLATE 36. The f for each score value is indicated by a dot. The f for score value 86 is 4, so the dot for this score value is placed directly above score value 86 and directly _____ from f 4.

■ ■ ■ ■ ■ ■ ■ ■ ■ ■ ■ ■ ■ ■

across

---

19. PLATE 36. When the dots that represent the f's of each score value are connected by a line, this line gives the "shape" of the distribution. The mode is score value _____ because it has the highest peak in the polygon.

■ ■ ■ ■ ■ ■ ■ ■ ■ ■ ■ ■ ■

86

---

20. Plate 37 presents a frequency polygon for a frequency distribution of grouped data. For grouped data, the designation along the horizontal axis is for the _____ of each interval instead of for each X as in ungrouped data.

Plate 37.

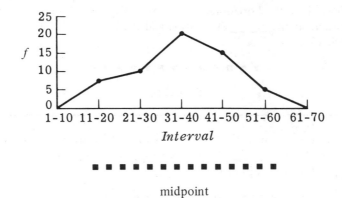

midpoint

---

21. PLATE 37. The vertical axis is marked off in steps of __ f's each. Where the f for a particular score value is between the f markings on the vertical axis, the dot representing the f for that score value is placed _____ the markings.

5, between

---

22. PLATE 37. The f for interval 11-20 is probably ___ because it lies about halfway between the f markings of 5 and 10.

7

---

23. PLATE 37. The f for interval 21-30 is ____ .

10

Plate 38.

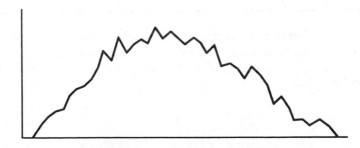

72

24. Plate 38 presents a frequency polygon in which there are frequencies depicted for many score values. When there are many different score values involved in a frequency polygon, and the N is quite large, the "trend" of the distribution ___ (is/is not) apparent.

■ ■ ■ ■ ■ ■ ■ ■ ■ ■ ■ ■ ■ ■ ■

is

Plate 39.

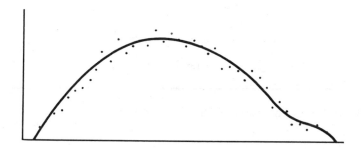

25. Plate 39 presents the "smooth" curve for the frequency polygon presented in Plate 38. As shown here, such smooth curves are generally obtained by drawing the best fitting _____ line through the many points rather than by connecting them with _____ lines.

■ ■ ■ ■ ■ ■ ■ ■ ■ ■ ■ ■ ■ ■ ■

curved, straight

---

26. In this way, a smooth curve approximates the frequency polygon. A number of types of curves can be obtained, depending on the way the scores are distributed.

Plate 40. A Curve of a Bimodal Distribution

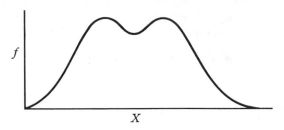

This curve is for a *bimodal* distribution because two score values have the same frequency, which indicates that there are two _____ in the distribution.

■ ■ ■ ■ ■ ■ ■ ■ ■ ■ ■ ■ ■ ■

modes

Plate 41. A Curve of a Negatively Skewed Distribution

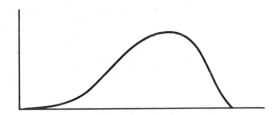

27. Another type of curve is depicted in Plate 41. This curve is said to be *skewed to the left* because the slope of the curve trails off to the _____. This is sometimes referred to simply as a *skewed distribution.*

■ ■ ■ ■ ■ ■ ■ ■ ■ ■ ■ ■ ■ ■

left

---

28. PLATE 41. This curve, which is skewed to the left, is said to be *negatively skewed* since the "tail" of the curve trails off to the end of the horizontal axis that represents the _____ (lower/higher) score values.

■ ■ ■ ■ ■ ■ ■ ■ ■ ■ ■ ■ ■ ■

lower

Plate 42. A Curve of a Positively Skewed Distribution

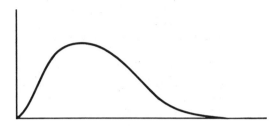

29. This is also a *skewed distribution*. The frequency polygon is skewed to the _____ because the slope of the curve trails off to the _____ .

■ ■ ■ ■ ■ ■ ■ ■ ■ ■ ■ ■ ■

right, right

---

30. PLATE 42. This curve is _____ (positively/negatively) skewed because the slope of the curve trails off to the right.

■ ■ ■ ■ ■ ■ ■ ■ ■ ■ ■ ■ ■

positively

---

31. A good way to determine whether a curve is negatively or positively skewed is to remember that if there were negative score values in the distribution they would be at the _____ (left/right) end of the horizontal axis.

■ ■ ■ ■ ■ ■ ■ ■ ■ ■ ■ ■ ■

left

Plate 43. A Curve of a Symmetrical Distribution

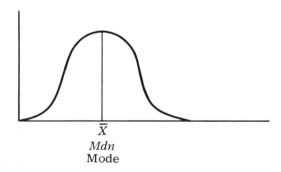

$\overline{X}$
Mdn
Mode

32. Plate 43 depicts a frequency polygon that is symmetrical. That is, if the curve is divided in half by a line drawn perpendicular to the horizontal axis, the two halves are identical in shape. This is called a _____ distribution.

■ ■ ■ ■ ■ ■ ■ ■ ■ ■ ■ ■ ■

symmetrical

---

33. PLATE 43. The perpendicular line marks the peak of the distribution; therefore, it designates the _____(measure of central tendency) of the distribution.

■ ■ ■ ■ ■ ■ ■ ■ ■ ■ ■ ■ ■

mode

---

34. The distributions shown in Plates 41, 42, and 43 are all called *unimodal* distributions because they all have only_____ mode.

■ ■ ■ ■ ■ ■ ■ ■ ■ ■ ■ ■ ■

one

---

35. In all unimodal symmetrical distributions, such as in Plate 43, the three measures of central tendency—the mode, the_____ (symbol), and the___ (symbol)—all coincide at the same score value.

■ ■ ■ ■ ■ ■ ■ ■ ■ ■ ■ ■ ■

Mdn, X̄

Plate 44. A Curve of a Positively Skewed Distribution

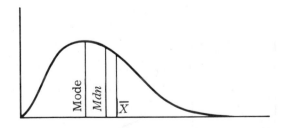

36. Plate 44 presents a frequency polygon of a positively skewed distribution. When a distribution is skewed, the mean, median, and the mode _____ (do/do not) coincide.

■ ■ ■ ■ ■ ■ ■ ■ ■ ■ ■ ■ ■ ■

do not

---

37. PLATE 44. The mode, of course, occurs at the peak of the curve. Because the _____ is affected by the extreme score values that occur at the *tail* of a skewed distribution, it will always fall somewhere between the mode and _____ of the distribution.

■ ■ ■ ■ ■ ■ ■ ■ ■ ■ ■ ■ ■ ■

mean, tail

---

38. PLATE 44. In a skewed distribution, the *Mdn* is not as affected by extreme scores as the $\bar{X}$; therefore, it lies somewhere between the _____ and the _____ of the distribution.

■ ■ ■ ■ ■ ■ ■ ■ ■ ■ ■ ■ ■ ■

mode, mean

---

39. To summarize, the order of the three measures of central tendency for a unimodal skewed distribution, starting from the peak and proceeding toward the tail, is _____ , _____ , and _____ .

■ ■ ■ ■ ■ ■ ■ ■ ■ ■ ■ ■ ■ ■

mode, median, mean

EXERCISES

1. Draw bar histograms for the following frequency distributions.

(a)

X	f
20	2
19	4
18	4
17	7
16	6
15	3
14	2
13	2

(b)

interval	f
36-40	2
31-35	5
26-30	4
21-25	3
16-20	5
11-15	2
6-10	1

2. Draw frequency polygons for the following frequency distributions.

(a)

X	f
106	1
105	0
104	3
103	4
102	5
101	7
100	5

(b)

interval	f
105-108	4
101-104	6
97-100	9
93-96	10
89-92	9
85-88	6
81-84	4

3. Which distribution(s) in Exercises 1 and 2 can be termed unimodal distribution(s)?

4. Which can be termed bimodal distribution(s)?

5. Which can be termed skewed distribution(s)? Is the skewness to the left or to the right?

6. Which can be termed symmetrical distribution(s)?

7. Which measure of central tendency is most affected by the degree of skewness in a frequency distribution? Which is least affected?

MEASURES OF VARIABILITY: RANGE, SEMI-INTERQUARTILE RANGE, AVERAGE DEVIATION

A bar histogram, frequency polygon, or "smoothed" curve gives us a graphic representation of a frequency distribution but does not provide us with a quantitative way of describing the spread of scores. In addition to the measure of central tendency, we need a measure of variability that describes the dispersion of the raw scores. In this set there will be a discussion of the concept of variability and a definition of range, interquartile range, and semi-interquartile range. These measures give some indication of the spread, or variability, of scores. However, they are usually too gross for the statistician, who needs a more refined measure of variability—one that takes into account the actual value of each of the raw scores. This set will present the meaning of an absolute deviation score and the method of calculating the average deviation. It will compare the average deviation with the range and semi-interquartile range.

SPECIFIC OBJECTIVES OF SET 6

At the conclusion of this set, you will be able to:

(1) calculate the range of a frequency distribution using Formula 4.

(2) calculate the interquartile range in a frequency distribution.

(3) calculate the semi-interquartile range in a frequency distribution using Formula 5.

(4) calculate positive and negative deviation scores, using Formula 6.

(5) calculate the average deviation of a distribution, using Formula 7.

(6) identify the symbols x and $|x|$.

1. To say that a group of data has a $\bar{X}$ of 19 or a *Mdn* of 27 gives you a representative score for the data but tells you nothing about the *spread* of the scores, or the extent to which they *vary*. To describe a set of scores fully, you need both a representative score and a measure of the extent to which they _____.

■ ■ ■ ■ ■ ■ ■ ■ ■ ■ ■ ■ ■ ■

vary

2. Two sets of data may have the same $\bar{X}$, but the spread—or variability—of the raw scores around the $\bar{X}$ may be quite different.

Plate 45.

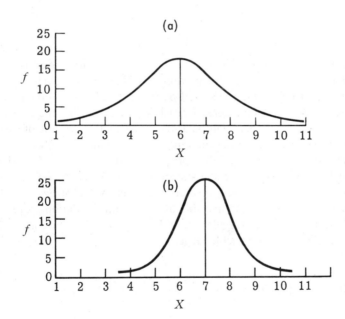

By inspection of the curves in Plate 43, it is evident that the scores in (a) are _____ (more/less) variable than those in (b).

■ ■ ■ ■ ■ ■ ■ ■ ■ ■ ■ ■ ■ ■

more

3. The simplest measure of variability of a group of scores is the range.

Formula 4. Calculation of the Range

$$\text{Range} = H - L$$

in which H is the highest score in the frequency distribution

L is the lowest score in the frequency distribution

Using Formula 4, the range of a frequency distribution having 10 as the lowest score value and 25 as the highest score value is _____ .

■ ■ ■ ■ ■ ■ ■ ■ ■ ■ ■ ■ ■

$$25 - 10 = 15$$

---

4 PLATE 32, page 67. The range for this frequency distribution of grouped data is _____ .

■ ■ ■ ■ ■ ■ ■ ■ ■ ■ ■ ■ ■

$$45 - 11 = 34$$

---

5. The range is not a very useful measure because it tells you nothing about the distribution of the raw scores. It tells you only the spread of score values within which all the _____ _____ lie.

■ ■ ■ ■ ■ ■ ■ ■ ■ ■ ■ ■ ■

raw scores

---

6. More useful measures of variability are the *interquartile* range and the *semi-interquartile* range. The interquartile range is determined by the difference between Q_1 and Q_3 which encompasses ____ % of the scores.

■ ■ ■ ■ ■ ■ ■ ■ ■ ■ ■ ■ ■

50

---

7. Because the prefix "semi-" means half, in order to determine the *semi-interquartile* range of a frequency distribution, you must divide the *interquartile* range by__.

■ ■ ■ ■ ■ ■ ■ ■ ■ ■ ■ ■ ■

2

Formula 5. Calculation of the Semi-interquartile Range

$$Q = \frac{Q_3 - Q_1}{2}$$

---

8. Formula 5 presents the formula for the calculation of the semi-interquartile range. The symbol commonly used for this measure is Q. The Q for a distribution in which $Q_1 = 12$ and $Q_3 = 28$ is ___.

■ ■ ■ ■ ■ ■ ■ ■ ■ ■ ■ ■ ■

8

---

9. When describing a frequency distribution, Q is often reported along with the _____ (symbol for a measure of central tendency) because both of these measures are derived from score values that lie at percentiles in the distribution.

■ ■ ■ ■ ■ ■ ■ ■ ■ ■ ■ ■ ■

Mdn

---

10. If you are told that *Mdn* = 25, and $Q = 4$, you would know that ____% of the scores lie above and below score 25, and that the middle ____% of the scores lie within a range of 8 score points.

■ ■ ■ ■ ■ ■ ■ ■ ■ ■ ■ ■ ■

50, 50

---

11. You are given this information: lowest score is 105; highest score is 135; *Mdn* = 120; $Q = 7$. Range = ____. 50% of the scores lie between 105 and _____. 50% of the scores lie between _____ and 135. The middle 50% of the scores lie within a spread of ____ score points.

■ ■ ■ ■ ■ ■ ■ ■ ■ ■ ■ ■ ■

30, 120, 120, 14

82

12. You are given this information: $Q_1 = 30$; $Q_2 = 59$; $Q_3 = 85$.____% of the scores lie below the score of 85. The midpoint of the distribution is at score ____. $Q =$ _____.

■ ■ ■ ■ ■ ■ ■ ■ ■ ■ ■ ■ ■ ■

75, 59, 27.5

---

13. The range and Q do not take into consideration the value of all of the raw scores in a distribution. The range considers only the upper and lower scores. Q considers only the middle ____ % of the scores.

■ ■ ■ ■ ■ ■ ■ ■ ■ ■ ■ ■ ■ ■

50

---

14. If the uppermost score in any distribution is increased by 10 points, the range _____ (is increased/is decreased/remains the same). Q _____ (is increased/is decreased/remains the same).

■ ■ ■ ■ ■ ■ ■ ■ ■ ■ ■ ■ ■ ■

is increased, remains the same

---

15. A more useful measure of variability than either the range or semi-interquartile range to indicate how a set of raw scores vary from the mean is called the *average deviation*. This is simply the _____ amount that the raw scores deviate from the mean.

■ ■ ■ ■ ■ ■ ■ ■ ■ ■ ■ ■ ■ ■

average

---

16. To determine the *average deviation* of the raw scores from the mean, you must first determine the extent to which each raw score _____ from the mean.

■ ■ ■ ■ ■ ■ ■ ■ ■ ■ ■ ■ ■ ■

deviates

---

17. The amount that a raw score deviates from the mean is the difference between the value of the raw score and the value of the _____ .

■ ■ ■ ■ ■ ■ ■ ■ ■ ■ ■ ■ ■ ■

mean

---

18. The symbol x represents the amount of deviation of a raw score from the mean.

Formula 6. Calculation of a Deviation Score
$$x = X - \bar{X}$$

The formula for the deviation of one X from the $\bar{X}$ is presented in Formula 6, in which X is the value of the _____ and $\bar{X}$ is the value of the _____ .

■ ■ ■ ■ ■ ■ ■ ■ ■ ■ ■ ■ ■ ■

raw score, mean

---

19. FORMULA 6. If $X = 17$ and $\bar{X} = 14$, then $x = 3$ because $x = X - \bar{X}$. If $X = 20$ and $\bar{X} = 15$, then $x =$ ___.

■ ■ ■ ■ ■ ■ ■ ■ ■ ■ ■ ■ ■ ■ ■

5

---

20. FORMULA 6. If $X = 20$ and $\bar{X} = 25$, then $x = -5$ because $x = 20 - 25 = -5$. If $X = 70$ and $\bar{X} = 80$, then $x =$ _____ .

■ ■ ■ ■ ■ ■ ■ ■ ■ ■ ■ ■ ■ ■

−10

---

21. x is called a "deviation score" because it represents the amount of _____ of a particular X from the $\bar{X}$.

■ ■ ■ ■ ■ ■ ■ ■ ■ ■ ■ ■ ■ ■

deviation

---

22. Using Formula 6, the deviation scores for the raw scores that are smaller than $\bar{X}$ will be _____ (positive/negative). The deviation scores for the raw scores that are larger than $\bar{X}$ will be _____ (positive/negative).

■ ■ ■ ■ ■ ■ ■ ■ ■ ■ ■ ■ ■ ■

negative, positive

---

23. Because the $\bar{X}$ is the arithmetic average of all the raw scores in a distribution, it follows that the sum of the *negative* deviation scores is _____ to the sum of the *positive* deviation scores.

■ ■ ■ ■ ■ ■ ■ ■ ■ ■ ■ ■ ■ ■

equal

---

24. If the sum of the negative deviation scores is equal to the sum of the positive deviation scores, then the sum of all the deviation scores must be _____ .

■ ■ ■ ■ ■ ■ ■ ■ ■ ■ ■ ■ ■ ■

zero

---

25. If the sum of the negative deviation scores is equal to the sum of the positive deviation scores, the average of the sum of all deviation scores in a distribution is also equal to _____ .

■ ■ ■ ■ ■ ■ ■ ■ ■ ■ ■ ■ ■ ■

zero

---

26. When you do not consider the sign of a deviation score, but only its value, it is called an *absolute* deviation (A.D.) score. The average deviation is the sum of the *absolute* deviation scores divided by____(symbol).

■ ■ ■ ■ ■ ■ ■ ■ ■ ■ ■ ■ ■ ■

N

27. When you do not consider the sign of a deviation score, but only its value, it is called an _____ deviation score.

■ ■ ■ ■ ■ ■ ■ ■ ■ ■ ■ ■ ■ ■

absolute

---

28. The symbol for the absolute deviation score is $|x|$. If $X = 15$ and $\bar{X} = 10$, then $|x| = 5$. If $X = 45$ and $\bar{X} = 41$, then $|x| = \underline{\quad}$.

■ ■ ■ ■ ■ ■ ■ ■ ■ ■ ■ ■ ■ ■

4

Formula 7. Calculation of the Average Deviation

$$A.D. = \frac{\Sigma f |x|}{N}$$

in which $| x |$ is the absolute deviation of a raw score from the mean

29. Formula 7 presents the formula for the calculation of the average deviation. The symbol $\Sigma f |x|$ in this formula means that you sum the _____ deviation scores.

■ ■ ■ ■ ■ ■ ■ ■ ■ ■ ■ ■ ■ ■

absolute

Plate 46.

X	f	fX	x	$|x|$	$f|x|$
12	1	12	3	3	3
11	2	22	2	2	4
10	4	40	1	1	4
9	5	45	0	0	0
8	2	16	−1	1	2
7	3	21	−2	2	6
6	1	6	−3	3	3
	$N = \overline{18}$	$\Sigma fX = \overline{162}$			$\Sigma f|x| = \overline{22}$

$$\bar{X} = 9$$

30. Plate 46 presents a frequency distribution. The fourth column presents the deviation score for each score value. The fifth column presents the *absolute* deviation score values, which are the same as the deviation score values except that they are all _____ (positive/negative).

■ ■ ■ ■ ■ ■ ■ ■ ■ ■ ■ ■ ■ ■

positive

---

31. PLATE 46. $\bar{X}$ = 9. Using Formula 6, for the score value of 12 the deviation score value is:

$$x = X - \bar{X} = 12 - 9 = 3$$

The deviation score values appear in the fourth column of the plate. The deviation score value of score value 7 is ___.

■ ■ ■ ■ ■ ■ ■ ■ ■ ■ ■ ■ ■ ■

−2

---

32. PLATE 46. The absolute deviation score values that appear in the fifth column are the same as those in the ___(symbol) column except that the negative signs have been omitted.

■ ■ ■ ■ ■ ■ ■ ■ ■ ■ ■ ■ ■ ■

x

---

33. PLATE 46. The $f|x|$ for each score value is computed by multiplying the absolute deviation score value $|x|$ by its ___(symbol).

■ ■ ■ ■ ■ ■ ■ ■ ■ ■ ■ ■ ■ ■

f

---

34. PLATE 46. $\Sigma f|x|$ is obtained by_____the $f|x|$ for all score values.

■ ■ ■ ■ ■ ■ ■ ■ ■ ■ ■ ■ ■ ■

summing

---

35. PLATE 46. $\Sigma f |x| = 22, N = 18$.
    Using Formula 7 for the calculation of the average deviation, substitute these values for the symbols and determine A.D.
    Note: We call the average absolute deviation simply the average deviation, since the average of the simple deviations is always zero.

    ■ ■ ■ ■ ■ ■ ■ ■ ■ ■ ■ ■ ■ ■

    $$\text{A.D.} = \frac{22}{18} = 1.22$$

---

36. PLATE 46. The average deviation = 1.22. Therefore, the average number of score points that the raw scores in the distribution deviate from the _____ is 1.22.

    ■ ■ ■ ■ ■ ■ ■ ■ ■ ■ ■ ■ ■ ■

    mean

---

37. The average deviation takes into account the numerical values of the raw scores, whereas the range or semi-interquartile range does not. Therefore, the average deviation is a _____ (better/worse) indicator of the variability of the raw scores.

    ■ ■ ■ ■ ■ ■ ■ ■ ■ ■ ■ ■ ■ ■

    better

---

38. Using Plate 46 as a guide, determine the average deviation (A.D.) of the frequency distribution presented below.

X	f	fX	x	|x|	f|x|
90	3				
89	3				
88	1				
87	4				
86	1				
85	1				

■ ■ ■ ■ ■ ■ ■ ■ ■ ■ ■ ■ ■ ■

Plate 47.

X	f	fX	x	|x|	f|x|
90	3	270	2	2	6
89	3	267	1	1	3
88	1	88	0	0	0
87	4	348	−1	1	4
86	1	86	−2	2	2
85	1	85	−3	3	3
	$N = 13$	$\Sigma fX = 1144$			$\Sigma f\|x\| = 18$

$$\bar{X} = 88 \qquad\qquad \text{A.D.} = \frac{18}{13} = 1.4$$

---

39. Because the A.D. gives you only a measure of how much, on the average, the raw scores in a distribution deviate from the _____ , it is not as useful as the next measure of variability to be considered.

■ ■ ■ ■ ■ ■ ■ ■ ■ ■ ■ ■

mean

EXERCISES

1. Calculate the range using Formula 4, and the semi-interquartile range using Formula 5, for the following frequency distributions.

(a)

X	f
20	2
19	4
18	4
17	7
16	6
15	3
14	2
13	2

(b)

interval	f
36-40	2
31-35	5
26-30	4
21-25	3
16-20	5
11-15	2
6-10	1

(c)

X	f
106	1
105	0
104	3
103	4
102	5
101	7
100	5

(d)

interval	f
105-108	4
101-104	6
97-100	9
93-96	10
89-92	9
85-88	6
81-84	4

2. (a) If the highest score in a frequency distribution is reduced in value, but still remains the highest score, will the range of the distribution be affected? If so, how?
   (b) Will the semi-interquartile range be affected? If so, how?

3. What do the symbols x and $|x|$ represent?

4. What does the term "absolute deviation score" mean? How is it calculated?

5. For the following frequency distributions, calculate the positive and negative deviation scores, using Formula 6. Calculate the average deviation for each distribution, using Formula 7.

(a)

X	f
10	1
9	2
8	5
7	4
6	3
5	3
4	1

(b)

X	f
51	1
50	1
49	1
48	4
47	3
46	2
45	2

MEASURES OF VARIABILITY: VARIANCE, STANDARD DEVIATION

The average deviation, as presented in the last set, provides a measure of variability that has limited value, particularly in its usefulness when we are in the process of making statistical inferences. Two important measures of variability are introduced in this set—the *variance* and the *standard deviation*. These two measures will be used throughout most of the remainder of this text. We will illustrate two methods for computing the sample variance: the deviation score and the raw score method. In addition, the computation of the sample standard deviation will be presented.

The degree to which any raw score deviates from the mean may be expressed in terms of its standard deviation from the mean. Thus, the raw score can be expressed as a "relative deviate." This term is defined in this set, and we will show how it is calculated.

SPECIFIC OBJECTIVES OF SET 7

At the conclusion of this set you will be able to:

(1) use Formula 8a to calculate the sample variance of a frequency distribution by the deviation score method.

(2) use Formula 8b to calculate the sample variance of a frequency distribution by the raw score method.

(3) use Formula 9 to compute the sample standard deviation.

(4) use Formula 10 to calculate the "relative deviate" for specific scores in a frequency distribution.

(5) identify the symbols s'^2, s', and z.

1. In the calculation of the average deviation you needed to use the absolute deviation scores; that is, it was necessary to ignore the _____ of the deviation scores.

■ ■ ■ ■ ■ ■ ■ ■ ■ ■ ■ ■ ■ ■

sign

2. A mathematically more useful measure of variability, the *variance*, requires that the deviation scores be *squared*; thus we can take into account the sign of each deviation score.

Formula 8a.   Calculation of the Sample Variance

$$s'^2 = \frac{\Sigma f x^2}{N}$$

Formula 8a presents the method for the calculation of the variance by the deviation score method. You will notice that x is used in the formula. Recall from Formula 6 that x represents a _____ score. Do not confuse this with X, which represents a raw score.

■ ■ ■ ■ ■ ■ ■ ■ ■ ■ ■ ■ ■ ■

deviation

3. FORMULA 8a.   The expression $\Sigma f x^2$ means that in a frequency distribution you first square the deviation score value, then multiply it by the f of the score value, and then _____ all of these products.

■ ■ ■ ■ ■ ■ ■ ■ ■ ■ ■ ■ ■ ■

sum

4. FORMULA 8a.   Be sure that you square the deviation score value *before* you multiply by f and sum the products. Do *NOT* sum them first and then square the sum. After determining $\Sigma f x^2$, you divide it by ___ (symbol).

■ ■ ■ ■ ■ ■ ■ ■ ■ ■ ■ ■ ■ ■

N

Plate 48.

X	f	fX	x	x^2	fx^2
12	1	12	3	9	9
11	2	22	2	4	8
10	4	40	1	1	4
9	5	45	0	0	0
8	2	16	−1	1	2
7	3	21	−2	4	12
6	1	6	−3	9	9
$N = 18$		$\Sigma fX = 162$			$\Sigma fx^2 = 44$

$$\bar{X} = \frac{162}{18} = 9 \quad s'^2 = \frac{\Sigma fx^2}{N} = \frac{44}{18} = 2.44$$

5. Plate 48 presents a frequency distribution, with the calculation of its variance. For score value 12 the deviation score (x) is 3. To obtain x^2 you square the value of 3, which equals ___. The value of the deviation score squared appears in the x^2 column.

■ ■ ■ ■ ■ ■ ■ ■ ■ ■ ■ ■ ■

9

6. PLATE 48. For score value 12, $x = 3$ and $x^2 = 9$. The f for this score value is ___; therefore, to obtain fx^2 you must multiply 9 times ___, which equals ___. This value appears in the fx^2 column of the plate.

■ ■ ■ ■ ■ ■ ■ ■ ■ ■ ■ ■ ■

1, 1, 9

7. PLATE 48. For score value 7, $x = -2$. To obtain x^2 you must square -2, which equals ___. Remember, when you multiply a negative number times a negative number, you obtain a positive product.

■ ■ ■ ■ ■ ■ ■ ■ ■ ■ ■ ■ ■

4

8. PLATE 48. For score value 7, $x = -2$ and $x^2 = 4$. The f for this score value is 3; therefore, to obtain fx^2 you must multiply ___ times ___, which equals ____. This value appears in the fx^2 column of the plate.

■ ■ ■ ■ ■ ■ ■ ■ ■ ■ ■ ■ ■ ■

3, 4, 12

9. PLATE 48. Σfx^2 is determined by _____ all of the numbers in the fx^2 column.

■ ■ ■ ■ ■ ■ ■ ■ ■ ■ ■ ■ ■ ■

summing

10. PLATE 48. Using Formula 8, the s'^2 is calculated by substituting the numerical values for the symbols and doing the necessary arithmetic. For this frequency distribution, $s'^2 = $ _____ .

■ ■ ■ ■ ■ ■ ■ ■ ■ ■ ■ ■ ■ ■

2.44

Plate 49.

X	f
21	1
20	2
19	3
18	5
17	3
16	2
15	1

11. Determine the $\bar{X}$ and s'^2 of the frequency distribution. Use Plate 48 as a guide.

$$\bar{X} = \underline{\hspace{3cm}} , \; s'^2 = \underline{\hspace{3cm}} \text{ (to two decimal places)}$$

■ ■ ■ ■ ■ ■ ■ ■ ■ ■ ■ ■ ■ ■

Plate 50.

X	f	fX	x	x^2	fx^2
21	1	21	3	9	9
20	2	40	2	4	8
19	3	57	1	1	3
18	5	90	0	0	0
17	3	51	−1	1	3
16	2	32	−2	4	8
15	1	15	−3	9	9
$N = 17$		$\Sigma fX = 306$			$\Sigma fx^2 = 40$

$$\bar{X} = \frac{306}{17} = 18 \qquad s'^2 = \frac{40}{17} = 2.35$$

12. When there is a large N and many score values involved in a distribution, it is very laborious to compute the s'^2 by the deviation score method. Another method of computing the s'^2 is the *raw score* method. This method does not use deviation scores. Instead, it uses the _____ scores in its computation.

■ ■ ■ ■ ■ ■ ■ ■ ■ ■ ■ ■ ■ ■ ■

raw

Formula 8b. Calculation of the Sample Variance

$$s'^2 = \frac{\Sigma fX^2}{N} - \bar{X}^2$$

13. Formula 8b presents the formula for the computation of s'^2 by the raw score method. In this formula, the symbol ΣfX^2 indicates that you must first square the _____ _____ value; then multiply by its f; then _____ all of the fX^2 values.

■ ■ ■ ■ ■ ■ ■ ■ ■ ■ ■ ■ ■ ■ ■

raw score, sum

---

14. FORMULA 8b. After obtaining ΣfX^2 you divide it by ___ (symbol) and then subtract the square of the ___ (symbol).

■ ■ ■ ■ ■ ■ ■ ■ ■ ■ ■ ■ ■ ■ ■

$N, \bar{X}$

Plate 51.

X	f	fX	X^2	fX^2
12	1	12	144	144
11	2	22	121	242
10	4	40	100	400
9	5	45	81	405
8	2	16	64	128
7	3	21	49	147
6	1	6	36	36
	$N = 18$	$\Sigma fX = 162$		$\Sigma fX^2 = 1502$

$$\bar{X} = \frac{162}{18} = 9 \quad s'^2 = \frac{1502}{18} - (9)^2 = 2.44$$

15. Plate 51 presents a frequency distribution and the calculation of s'^2 by the raw-score method using Formula 8b. Notice that each score value is squared *before* it is multiplied by f, and then these fX^2 values are _____ to obtain ΣfX^2.

■ ■ ■ ■ ■ ■ ■ ■ ■ ■ ■ ■ ■ ■

summed

---

16. PLATE 51. Each f is multiplied by the square of the score value, and these are summed to provide ΣfX^2. This figure is divided by ___ (symbol). From this dividend, the square of the ___ (symbol) is subtracted.

■ ■ ■ ■ ■ ■ ■ ■ ■ ■ ■ ■ ■ ■

$N, \bar{X}$

---

17. Plates 48 and 51 have the same frequency distribution. The s'^2 obtained in Plate 51 by the _____ score method is exactly the same as that obtained in Plate 48 by the _____ score method.

■ ■ ■ ■ ■ ■ ■ ■ ■ ■ ■ ■ ■ ■

raw, deviation

---

18. Determine the $\bar{X}$ and s'^2 of the frequency distribution below. Use Formula 8b for the calculation of s'^2 by the raw score method. Use Plate 51 as a guide.

X	f
21	1
20	2
19	3
18	5
17	3
16	2
15	1

$\bar{X} =$ _____ , $s'^2 =$ _____ (to two decimal places)

■ ■ ■ ■ ■ ■ ■ ■ ■ ■ ■ ■ ■ ■

Plate 52.

X	f	fX	X^2	fX^2
21	1	21	441	441
20	2	40	400	800
19	3	57	361	1083
18	5	90	324	1620
17	3	51	289	867
16	2	32	256	512
15	1	15	225	225
	$N = 17$	$\Sigma fX = 306$		$\Sigma fX^2 = 5548$

$$\bar{X} = \frac{306}{17} = 18 \quad s'^2 = \frac{5548}{17} - (18)^2 = 2.35$$

---

19. Although the sum of squares and the variance are important measures of variability (which shall be used later in connection with statistical inference), in describing a set of data another measure, the *standard deviation*, is useful.

Formula 9. Calculation of the Sample Standard Deviation

$$s' = \sqrt{s'^2}$$

As depicted in Formula 9, the standard deviation of a set of data is the _____ _____ of the variance.

■ ■ ■ ■ ■ ■ ■ ■ ■ ■ ■ ■ ■

square root

---

20. The standard deviation (s') is usually considered the *standard* unit for describing the deviations of a set of raw scores from the ___ (symbol).

■ ■ ■ ■ ■ ■ ■ ■ ■ ■ ■ ■ ■

$\bar{X}$

---

21. Many times it is desirable to describe a deviation score (x) in terms of how many standard deviations (s') it represents. If for a particular raw score the deviation score is 10, this means that the raw score lies _____ points above the $\overline{X}$.

■ ■ ■ ■ ■ ■ ■ ■ ■ ■ ■ ■ ■

10

22. If the raw score lies 10 points above the $\overline{X}$, then $x = 10$. If the s' of the distribution is 5, this raw score is ___s' above the $\overline{X}$.

■ ■ ■ ■ ■ ■ ■ ■ ■ ■ ■ ■ ■

2

23. Similarly, if for a particular score $x = 12$ and $s' = 4$, then you know that the raw score lies 12/4 or _____ above the $\overline{X}$.

■ ■ ■ ■ ■ ■ ■ ■ ■ ■ ■ ■ ■

$3s'$

24. Any raw score can be described in terms of how much it deviates from the $\overline{X}$, in standard deviation units, by dividing its x by the ___(symbol) of the distribution.

■ ■ ■ ■ ■ ■ ■ ■ ■ ■ ■ ■ ■

s'

25. When you divide a deviation score (x) by the standard deviation of the distribution (s'), the dividend is termed a standard score, or a *relative deviate*.

Formula 10. Calculation of a Relative Deviate

$$z = \frac{x}{s'}$$

Formula 10 indicates that the standard score derived by this method is given the symbol___.

■ ■ ■ ■ ■ ■ ■ ■ ■ ■ ■ ■ ■ ■

z

---

26. Formula 10 is used in calculating z. This is called a *relative deviate* because it is a standardized method of describing the deviation of a value from its mean, _____ to the standard deviation of the values in the distribution.

■ ■ ■ ■ ■ ■ ■ ■ ■ ■ ■ ■ ■ ■

relative

---

27. The numerical value of a relative deviate (z) is called a z-score. Thus, if for a particular raw score $z = 1.7$, this is referred to as z-score_____.

■ ■ ■ ■ ■ ■ ■ ■ ■ ■ ■ ■ ■ ■

1.7

---

28. To use Formula 10 to determine the z-score for a particular X, you must first subtract the $\overline{X}$ from the X to determine ___ (symbol).

■ ■ ■ ■ ■ ■ ■ ■ ■ ■ ■ ■ ■ ■

x

---

29. You are given the following information about a frequency distribution: $\overline{X} = 50$, $s' = 10$. To determine the z-score for raw score 75, first determine x by using Formula 6. $x =$ _____.

■ ■ ■ ■ ■ ■ ■ ■ ■ ■ ■ ■ ■ ■

25 $(75 - 50)$

---

30. Same distribution: $\overline{X} = 50$, $s' = 10$. For raw score 75, $x = 25$. To determine the z-score for this raw score, substitute the numerical values for the symbols in Formula 10. $z =$ _____.

■ ■ ■ ■ ■ ■ ■ ■ ■ ■ ■ ■ ■ ■

2.5 $(25/10)$

---

31. Same distribution: $\overline{X}$ = 50, s' = 10. For raw score 75, the z-score is 2.5. This means that raw score lies _____ standard deviations above the mean, or at a point designated as _____ s'.

■ ■ ■ ■ ■ ■ ■ ■ ■ ■ ■ ■ ■ ■

2.5, 2.5

32. Same distribution: $\overline{X}$ = 50, s' = 10. For raw score 75 the z-score is 2.5, which is a positive number because, for this raw score x was positive. If, for a given raw score, x is negative, then its z-score will also be _____ .

■ ■ ■ ■ ■ ■ ■ ■ ■ ■ ■ ■ ■ ■

negative

33. Same distribution: $\overline{X}$ = 50, s' = 10. Determine the x and the z-score for the raw score 34. x = _____ , z = _____ .

■ ■ ■ ■ ■ ■ ■ ■ ■ ■ ■ ■ ■ ■

-16	$(34 - 50)$
-1.6	$(-16/10)$

34. Thus, the use of the z-score provides a standardized way of describing a score's relative _____ from the _____ . The use of z-scores will take on added usefulness in later sets.

■ ■ ■ ■ ■ ■ ■ ■ ■ ■ ■ ■ ■ ■

deviation, mean

EXERCISES

1. Use Formulas 8a and 9 to calculate the sample variance and the standard deviation of the following frequency distribution by the deviation score method.

X	f
10	1
9	2
8	3
7	4
6	5
5	4
4	3
3	2
2	1

2. Use Formulas 8b and 9 to calculate the sample variance and the standard deviation of the following frequency distribution by the raw score method.

X	f
20	1
19	4
18	6
17	3
16	2
15	1

3. What is meant by the term *relative deviate*? What is its symbol?

4. Determine the z-score for score 9 in Exercise 1; for score 16 in Exercise 2.

RELATIONSHIP BETWEEN POPULATION AND SAMPLE / INTRODUCTION TO THE NORMAL DISTRIBUTION

In research you are seldom afforded the luxury of obtaining a raw score for each person in a population. The common practice is to obtain a sample of individuals, determine raw scores for them, and apply statistical techniques to these sample scores.

Different notations are used in statistics to describe characteristics of samples and those of populations. In this set the terminology and symbolic notations generally used in statistical analysis will be presented.

This set will introduce you to the most important function of statistics—that of providing a basis upon which inferences can be made from a sample to a population. These terms will be defined, and a distinction made between the *mean* of a *population* and that of a *sample* and between the *variance* and *standard deviation* of a sample and that of a population.

In this set, you will also be introduced to a very important concept in statistics—that of the normal distribution. You will see the relationship between the mean, median, and mode in a normal distribution. In addition, you will see how the standard deviation relates to the area under the normal curve and how to determine the percentage of scores lying between various standard deviation values.

SPECIFIC OBJECTIVES FOR SET 8

At the conclusion of this set you will be able to:

(1) state the definition of *sample* and *population*.

(2) define the terms *parameter* and *statistic*.

(3) identify symbols which represent population parameters and sample statistics.

(4) state the relationship of the mean, median, and mode in a normal distribution.

(5) determine the percentage of scores lying between various standard deviation values.

(6) identify the symbols μ, σ^2 and σ.

1. When you measure the height of *all* of the seventh-grade children in New York City, you have measured the *population* of seventh-grade children in New York City. If you measure only fifty of these children, you have measured a *sample* of the _____.

■ ■ ■ ■ ■ ■ ■ ■ ■ ■ ■ ■ ■ ■

population

2. A population is defined by its descriptive terms. That is, all girls in high school English classes can be termed a population. All Boy Scouts, or all eight-day-old rats, can be termed a _____.

■ ■ ■ ■ ■ ■ ■ ■ ■ ■ ■ ■ ■ ■

population

3. A *sample* is any portion of the population less than the total population from which it is selected. Thus, if there are one hundred elementary school teachers in a population, fifteen of these would be a _____ of the population.

■ ■ ■ ■ ■ ■ ■ ■ ■ ■ ■ ■ ■ ■

sample

4. When you have measured all of the people or objects of a certain description, you have measured the population. When you have measured a selected group from the population, you have measured a _____ of the population.

■ ■ ■ ■ ■ ■ ■ ■ ■ ■ ■ ■ ■ ■

sample

5. Usually it is not possible to obtain measurements on every individual or object in the population. Therefore, you select a _____ and make the assumption that this _____ is representative of the population from which it is selected.

■ ■ ■ ■ ■ ■ ■ ■ ■ ■ ■ ■ ■ ■

sample, sample

6. It would be almost impossible to measure the height of every seventh-grade child in New York City, but you can measure the height of a sample of seventh-grade children and make inferences about the height of this _____ from the measurements obtained on the sample.

■ ■ ■ ■ ■ ■ ■ ■ ■ ■ ■ ■ ■ ■

population

---

7. If you wish to make inferences about the height of the population of seventh-grade children in New York City from measurements you obtain on a sample of that population, the sample must be *representative* of the _____ .

■ ■ ■ ■ ■ ■ ■ ■ ■ ■ ■ ■ ■ ■

population

---

8. A sample selected to represent as nearly as possible the population from which it is selected is called a representative sample. The most common method of obtaining a sample that is _____ of the population is by selecting it *randomly*.

■ ■ ■ ■ ■ ■ ■ ■ ■ ■ ■ ■ ■ ■

representative

---

9. When a sample is selected from a _____ by some random method, it is called a *random sample*.

■ ■ ■ ■ ■ ■ ■ ■ ■ ■ ■ ■ ■ ■

population

---

10. If you place the names of all seventh-grade children in New York City in a barrel, and draw out fifty names, you will have a _____ sample of this population.

■ ■ ■ ■ ■ ■ ■ ■ ■ ■ ■ ■ ■ ■

random

---

11. You may wish to know the mean height of the population of seventh-grade children in New York City, but you cannot measure the height of *all* children in this population. Instead, you can obtain the mean height of a _____ of this population.

■ ■ ■ ■ ■ ■ ■ ■ ■ ■ ■ ■ ■ ■

sample

12. From the mean score of a sample, you can make an inference about the mean score of the _____ of which the sample is representative.

■ ■ ■ ■ ■ ■ ■ ■ ■ ■ ■ ■ ■ ■

population

13. If you know the variance of scores for a sample, you can make an inference about the _____ of the scores in the population.

■ ■ ■ ■ ■ ■ ■ ■ ■ ■ ■ ■ ■ ■

variance

14. If your sample is a random sample of the seventh-grade children in New York City, you may make inferences from this sample to the _____ of seventh-grade children in New York City.

■ ■ ■ ■ ■ ■ ■ ■ ■ ■ ■ ■ ■ ■

population

15. You may *not* make inferences from this sample to all seventh-grade children in the United States, because your sample is not representative of this _____.

■ ■ ■ ■ ■ ■ ■ ■ ■ ■ ■ ■ ■ ■

population

16. Your sample permits you to make inferences only about the population from which it is _____.

■ ■ ■ ■ ■ ■ ■ ■ ■ ■ ■ ■ ■ ■

selected

---

17. The mean score for a sample is called a *sample mean*. The mean score for a population is called a _____ _____.

■ ■ ■ ■ ■ ■ ■ ■ ■ ■ ■ ■ ■ ■

population mean

---

18. You already know the symbol for a sample mean. It is ___ (symbol). The symbol for a population mean is μ.

■ ■ ■ ■ ■ ■ ■ ■ ■ ■ ■ ■ ■ ■

$\overline{X}$

---

19. When you refer to the mean of a population, you use ___ (symbol). When you refer to the mean of a sample, you use ___ (symbol).

■ ■ ■ ■ ■ ■ ■ ■ ■ ■ ■ ■ ■ ■

$\mu, \overline{X}$

---

20. You already know the symbol for the variance of a sample. It is _____ (symbol). The symbol for the variance of a population is σ^2.

■ ■ ■ ■ ■ ■ ■ ■ ■ ■ ■ ■ ■ ■

s'^2

---

21. Greek letters are used as symbols for the characteristics of populations. Roman letters are used as symbols for the characteristics of _____ .

■ ■ ■ ■ ■ ■ ■ ■ ■ ■ ■ ■ ■ ■

samples

---

22. The symbol for a population mean is ___. The symbol for the standard deviation of a population is ___ (symbol).

■ ■ ■ ■ ■ ■ ■ ■ ■ ■ ■ ■ ■

μ, σ

---

23. The symbol for a sample mean is ___ (symbol). The symbol for the standard deviation of a sample is ___ (symbol).

■ ■ ■ ■ ■ ■ ■ ■ ■ ■ ■ ■ ■

$\overline{X}, s'$

---

24. Characteristics of populations are represented by _____ (Greek/Roman) letters. Characteristics of samples are represented by (Greek/Roman) letters.

■ ■ ■ ■ ■ ■ ■ ■ ■ ■ ■ ■ ■

Greek, Roman

---

25. A characteristic of a population, such as its μ or σ, is called a *parameter*. The μ height of all seventh-grade children in New York City is called a

_____ .

■ ■ ■ ■ ■ ■ ■ ■ ■ ■ ■ ■ ■

parameter

---

26. A *parameter* is a characteristic of a population. A *statistic* is a characteristic of a sample. The μ is called a _____ . The $\overline{X}$ is called a

_____ .

■ ■ ■ ■ ■ ■ ■ ■ ■ ■ ■ ■ ■

parameter, statistic

Note: The words "population" and "parameter" both begin with the letter "p". The words "sample" and "statistic" both begin with the letter "s".

27. The σ of the height of the population of seventh-grade children in New York City is called a _____ (parameter/statistic).

■ ■ ■ ■ ■ ■ ■ ■ ■ ■ ■ ■ ■

parameter

---

28. The s' of a sample is called a _____ (parameter/statistic).

■ ■ ■ ■ ■ ■ ■ ■ ■ ■ ■ ■ ■

statistic

---

29. The mean of a sample is written ___ (symbol) and is called a _____ (parameter/statistic). The mean of a population is written ___ (symbol) and is called a _____ (parameter/statistic).

■ ■ ■ ■ ■ ■ ■ ■ ■ ■ ■ ■ ■

$\overline{X}$, statistic, μ, parameter

---

30. In all previous sections of this text we have dealt exclusively with _____ (Greek/Roman) symbols. Thus, as indicated by our symbols, we were concerned with describing distributions of _____ (sample/population) values.

■ ■ ■ ■ ■ ■ ■ ■ ■ ■ ■ ■ ■

Roman, sample

---

31. We are now in a position to consider the properties of a particular type of population distribution. Recall that as we add more and more scores to a distribution, its shape can often be depicted with increasing accuracy by means of a smoothed _____ .

■ ■ ■ ■ ■ ■ ■ ■ ■ ■ ■ ■ ■ ■

curve

32. One smoothed curve of particular value to statisticians is the one depicting the distribution of the frequency with which values occur *on the basis of chance alone*. This particular distribution, called the normal distribution, is depicted by a smoothed curve, appropriately called a _____ curve.

■ ■ ■ ■ ■ ■ ■ ■ ■ ■ ■ ■ ■ ■

normal

33. The normal curve is a theoretical concept developed from a mathematical equation, quite apart from any concrete data. Its important feature is that it accurately reflects the distribution of many sets of data, the values of which vary solely due to _____ .

■ ■ ■ ■ ■ ■ ■ ■ ■ ■ ■ ■ ■ ■

chance

Plate 53.

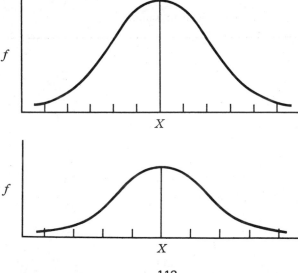

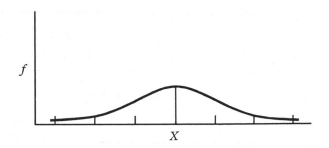

f

X

34. Plate 53 presents three normal curves. For each curve, the score values are indicated along the horizontal axis. The frequency of scores are indicated along the vertical axis. The exact shapes of these normal curves differ, depending upon the distance between_____ _____ along the horizontal axis.

■ ■ ■ ■ ■ ■ ■ ■ ■ ■ ■ ■ ■

score values

---

35. The curves depicted in Plate 53 indicate that, as the scale of score values along the horizontal axis differs, the specific shape of the normal curve changes. Thus, there _____(is/is not) one specific shape that represents the normal distribution.

■ ■ ■ ■ ■ ■ ■ ■ ■ ■ ■ ■ ■

is not

---

36. Plate 53 reveals that there is a family of normal curves, the exact shape depending upon the distance between values along the horizontal axis. However, as indicated by any of the curves in this plate, a normal curve _____ (is/is not) symmetrical around a central value.

■ ■ ■ ■ ■ ■ ■ ■ ■ ■ ■ ■ ■

is

---

37. Because of the bell-shaped symmetrical nature of the normal curve, its mean, median, and mode _____ (do/do not) coincide.

■ ■ ■ ■ ■ ■ ■ ■ ■ ■ ■ ■ ■

do

---

38. Because the theoretical normal curve is developed from a hypothetical infinite population of scores, it is proper to designate its mean as ___ $(\overline{X}/\mu)$ and its standard deviation as ___ (s'/σ).

■ ■ ■ ■ ■ ■ ■ ■ ■ ■ ■ ■ ■ ■

μ, σ

---

39. Since the normal curve is symmetric around its mean, μ, we know ___ % of the *total* area under the curve lies to the left of μ and ___ % lies to the right of μ.

■ ■ ■ ■ ■ ■ ■ ■ ■ ■ ■ ■ ■

50, 50

Plate 54.

$$-4\sigma \quad -3\sigma \quad -2\sigma \quad -1\sigma \quad 0 \quad 1\sigma \quad 2\sigma \quad 3\sigma \quad 4\sigma$$
$$\mu$$

40. To obtain a picture of the properties of the normal curve, in Plate 54 we have specified points along a horizontal axis as representing standard deviation units. The mean (μ) has been set at zero and the standard deviation (σ) units have been designated as specified distances along the axis. The point designating the value at one standard deviation to the right of the mean is given the symbol ___.

■ ■ ■ ■ ■ ■ ■ ■ ■ ■ ■ ■ ■

1σ

---

41. The distances between standard deviation designations along the axis are ___ (equal/unequal).

■ ■ ■ ■ ■ ■ ■ ■ ■ ■ ■ ■ ■

equal

The shape of the theoretical normal curve with $\mu = 0$ and $\sigma = 1$ has been developed by the use of a complex mathematical formula (which we need not examine here) and is presented in Plate 55.

Plate 55.

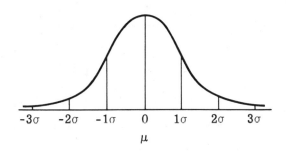

42. As with all smoothed distribution curves, the entire area under the curve is taken as representing _____ % of the scores in the distribution.

■ ■ ■ ■ ■ ■ ■ ■ ■ ■ ■ ■ ■

100

Because of its invariant shape, given specific distances for standard deviation units along the horizontal axis, mathematicians have determined the proportion of the total area contained in any segment of the area under the curve.

Plate 56. Normal Curve

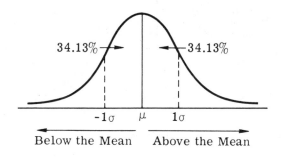

43. PLATE 56. The point designated by 1σ is said to be one standard deviation *above* the mean because it is located on the right of the μ where the score values are larger than the μ. The same point to the left of the μ is designated -1σ and is said to be one standard deviation _____ the μ.

■ ■ ■ ■ ■ ■ ■ ■ ■ ■ ■ ■ ■ ■

below

---

44. PLATE 56. In the normal curve, the distance between the μ and 1σ encompasses 34.13% of the area under the curve. Because the area under the curve represents the frequency of the scores in the distribution, it can be said that _____ % of the scores are contained between the μ and 1σ.

■ ■ ■ ■ ■ ■ ■ ■ ■ ■ ■ ■ ■ ■

34.13

---

45. PLATE 56. Because the distribution is symmetrical, the distance between the μ and -1σ also encompasses _____ % of the scores.

■ ■ ■ ■ ■ ■ ■ ■ ■ ■ ■ ■ ■ ■

34.13

Plate 57. Normal Curve

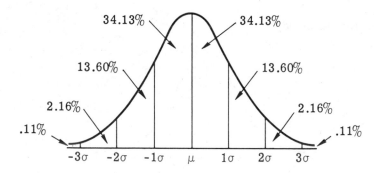

Note: Figures refer to percentages of total area bounded by a normal curve.

46. Plate 57 indicates how much of the area of the normal curve is included between any two perpendiculars indicating standard deviation units. The distance along the horizontal axis from μ to 1σ is equal to the distance between 1σ and 2σ. This is the same distance as between 2σ and ____ .

■ ■ ■ ■ ■ ■ ■ ■ ■ ■ ■ ■ ■

3σ

---

47. PLATE 57. The standard deviation designations to the left of the μ are equal to those on the right of the μ. The proportion of scores contained in corresponding areas to the left and right of μ ____ (are/are not) exactly the same.

■ ■ ■ ■ ■ ■ ■ ■ ■ ■ ■ ■ ■

are

---

48. PLATE 57. The percentage of scores that lie between μ and 1σ is 34.13%, but the percentage of scores that lie between 1σ and 2σ is only ____ %.

■ ■ ■ ■ ■ ■ ■ ■ ■ ■ ■ ■ ■

13.60

---

49. The percentage of scores that lie between μ and 2σ is obtained by summing the percentages contained between these two points: 34.13% + 13.60% = ____ %.

■ ■ ■ ■ ■ ■ ■ ■ ■ ■ ■ ■ ■

47.73

---

50. PLATE 57. Between -1σ and 1σ are contained ____ % of the scores.

■ ■ ■ ■ ■ ■ ■ ■ ■ ■ ■ ■ ■

68.26

---

51. PLATE 57. The percentage of scores that lies between -3σ and 3σ is ____ %.

■ ■ ■ ■ ■ ■ ■ ■ ■ ■ ■ ■ ■

99.78

---

52. The mathematically derived curve of a normal distribution never touches the horizontal axis. However, for practical purposes very few scores lie outside of −3σ and 3σ.

PLATE 57. _____ % of the scores lie outside of −3σ and 3σ.

■ ■ ■ ■ ■ ■ ■ ■ ■ ■ ■ ■ ■ ■

.22

---

53. Remember, the normal curve is not a frequency polygon of an actual set of data, but is a mathematically derived curve. The properties of the normal curve are very useful to statisticians because the frequency distributions of many sets of data resemble the _____ curve.

■ ■ ■ ■ ■ ■ ■ ■ ■ ■ ■ ■ ■ ■

normal

---

54. If you assume that a set of data is normally distributed, given its μ and σ, you can describe fully the set of data. For illustration, assume that a population of spelling scores for seventh-grade children is normally distributed with $\mu = 80$ and $\sigma = 5$. Thus, 1σ above the μ is 5 points above score value 80, which is score value ____ .

■ ■ ■ ■ ■ ■ ■ ■ ■ ■ ■ ■ ■ ■

85

Plate 58. Frequency Distribution of Spelling Scores

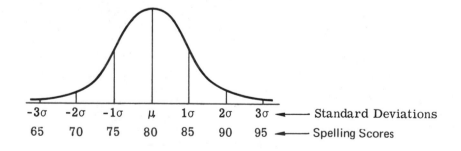

| | −3σ | −2σ | −1σ | μ | 1σ | 2σ | 3σ | ◄——— Standard Deviations |
| | 65 | 70 | 75 | 80 | 85 | 90 | 95 | ◄——— Spelling Scores |

55. Plate 58 presents the normal distribution of spelling scores where $\mu = 80$ and $\sigma = 5$. The score value of 80 is placed at the μ, and 1σ is 5 points *above* the μ, at score value 85.-1σ is 5 points *below* the μ at score value ___ .

■ ■ ■ ■ ■ ■ ■ ■ ■ ■ ■ ■ ■ ■

75

---

56. Refer to Plate 57 whenever you need the properties of the normal distribution.

PLATE 58. Between the μ and 1σ are_____% of the scores. Therefore, if the $\mu = 80$ and $\sigma = 5$, then _____ % of the children received scores between 80 and 85.

■ ■ ■ ■ ■ ■ ■ ■ ■ ■ ■ ■ ■ ■

34.13, 34.13

---

57. PLATE 58. 34.13% of the children received scores between the values of 80 and 85. _____% of the children received scores between the values of 85 and 90. _____% of the children received scores between the values of 80 and 90.

■ ■ ■ ■ ■ ■ ■ ■ ■ ■ ■ ■ ■ ■

13.60
47.73   (34.13% + 13.60%)

---

58. PLATE 58._____ % of the children received scores between the values of 75 and 85.

■ ■ ■ ■ ■ ■ ■ ■ ■ ■ ■ ■ ■

68.26   (34.13% + 34.13%)

---

59. PLATE 58. _____% of the children received scores between the values of 70 and 90.

■ ■ ■ ■ ■ ■ ■ ■ ■ ■ ■ ■ ■ ■

95.46
(13.60% + 34.13% + 34.13% + 13.60%)

---

60. PLATE 58. _____ % of the children received scores lower than 65.

■ ■ ■ ■ ■ ■ ■ ■ ■ ■ ■ ■ ■ ■

.11

---

61. PLATE 58. _____ % of the children received scores outside of score values 65 and 95.

■ ■ ■ ■ ■ ■ ■ ■ ■ ■ ■ ■ ■ ■

.22

---

62. PLATE 58. For convenience of computations, assume that the population contains only 300 seventh-grade children's spelling scores. If $N = 300$, the number of children receiving scores between the μ and 1σ would be 34.13% of 300, which is 102.4 scores. This means that a frequency of 102.4 scores lies between the score values of _____ and _____.

■ ■ ■ ■ ■ ■ ■ ■ ■ ■ ■ ■ ■ ■

80, 85

---

63. PLATE 58. If $N = 300$, the number of children that received scores between the values of 75 and 85 would be _____ .

■ ■ ■ ■ ■ ■ ■ ■ ■ ■ ■ ■ ■

205   (34.13% + 34.13% = 68.26%)
             (68.26% X 300 = 204.78,
             or approximately 205)

EXERCISES

1. What is meant by the terms *population* and *sample*?

2. What term is used to describe characteristics of a population? Of a sample?

3. What type of symbols are used to represent population characteristics? Of sample characteristics?

4. What is the position of the μ, *Mdn*, and mode in a normal distribution?

5. Using Plate 57 (page 116), answer the following questions.
   (a) What percentage of scores lies between μ and 1σ?
   (b) What percentage of scores lies between -1σ and 1σ?
   (c) What percentage of scores lies outside -2σ and 2σ?
   (d) What percentage of scores lies above 3σ?

6. What do the symbols μ, σ^2, and σ represent?

PROBABILITY AND THE
NORMAL DISTRIBUTION

This set extends the discussion of the standard deviation as it relates to the areas under the normal curve. Earlier you learned a method for converting a raw score into a "relative deviate" or z score. In this set, you will be introduced to the table of areas of the normal curve. This table will be used to determine the proportion of scores lying between the mean and various z scores.

Up to this point, all of the methods and concepts that have been introduced have been for the purpose of describing data. While the use of *descriptive statistics* fulfills an important role, the use of statistical techniques for making *inferences* is probably of greater value. In this set you will be introduced to probability as it relates to the normal curve. The concept of probability will concern us for the remainder of the text.

SPECIFIC OBJECTIVES OF SET 9

At the conclusion of this set you will be able to:

(1) express percentages as proportions.

(2) relate proportions under the normal curve to probability statements.

(3) use Table 1 to determine the probability values between the population mean and z scores for normally distributed data.

(4) identify the symbol P.

1. Recall that in describing deviations from the sample mean (x) in z-score units we used the formula x/s', where s' is the standard deviation of the

    _____.

    ■ ■ ■ ■ ■ ■ ■ ■ ■ ■ ■ ■ ■ ■

    sample

---

2. When dealing with population parameters, the deviation from the population mean is still designated x (now given by $X - \mu$) but the population standard deviation is represented by the symbol ___.

    Thus, for population data, $z = \dfrac{\overline{}}{\overline{}}$

    ■ ■ ■ ■ ■ ■ ■ ■ ■ ■ ■ ■ ■ ■

    $$\sigma, \frac{x}{\sigma}$$

---

3. In a distribution where $\mu = 0$ and $\sigma = 1$, the z-score at -1 standard deviations will be ___. The z-score at 3 standard deviations will be ___.

    ■ ■ ■ ■ ■ ■ ■ ■ ■ ■ ■ ■ ■

    $-1, 3$

---

4. It follows that whenever we set $\mu = 0$ and $\sigma = 1$ at any given point in the distribution, the z-score will correspond numerically to the _____ _____ value designating that point.

    ■ ■ ■ ■ ■ ■ ■ ■ ■ ■ ■ ■ ■ ■

    standard deviation

---

5. From Plate 57 we know that, in a standard normal curve where $\mu = 0$ and $\sigma = 1$, the area from μ to z-score 1 includes _____ % of the cases.

    ■ ■ ■ ■ ■ ■ ■ ■ ■ ■ ■ ■ ■

    34.13

Plate 59.

$\mu = 50$
$\sigma = 10$
$N = 300$

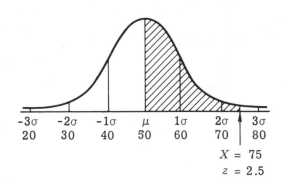

$X = 75$
$z = 2.5$

6. In this plate, for $X = 75$, $z = 2.5$. The position of z-score 2.5 is shown graphically. If you know that the z-score is 2.5, this is the same as saying that the raw score lies at a point _____ σ above the μ.

■ ■ ■ ■ ■ ■ ■ ■ ■ ■ ■ ■ ■ ■

2.5

7. From Plate 55, for normally distributed data, you know that _____ % of the scores lie between the μ and 2σ, which is the same as the percentage of scores that lies between μ and z-score 2. However, you cannot determine from Plate 55 what percentage of the scores lies between the μ and z-score 2.5.

■ ■ ■ ■ ■ ■ ■ ■ ■ ■ ■ ■ ■ ■

47.73

8. The various percentages lying between μ and any z-score value can be obtained from the use of statistical tables. These tables present the percentages in terms of proportions, which are the decimal equivalent of percentages. For example: 34.13% is expressed as .3413, 50% is expressed as .5, 49.38% is expressed as _____ .

■ ■ ■ ■ ■ ■ ■ ■ ■ ■ ■ ■ ■ ■

.4938

124

9. .65 is the decimal form of 65%. .5167 is the decimal form of 51.67%. .7216 is the decimal form of _____%.

■ ■ ■ ■ ■ ■ ■ ■ ■ ■ ■ ■ ■ ■ ■

72.16

Note: If you have not already done so, remove the Tables from the rear of the text and refer to them as directed.

10. Table 1 presents the areas of the standard normal curve that correspond to different values of z-scores. The first column of the table presents the different values of z-scores, and the second column presents the proportion of the total area under the curve that lies between the μ and the z-scores. The proportion for z-score 2.5 is _____.

■ ■ ■ ■ ■ ■ ■ ■ ■ ■ ■ ■ ■ ■ ■

.4938

11. PLATE 59. The area between the μ and z-score 2.5 is shaded. From Table 1, this represents .4938 of the area under the normal curve. Because the area is equivalent to the frequency of the scores in the distribution, it follows that _____ of N lie between the μ and z-score 2.5.

■ ■ ■ ■ ■ ■ ■ ■ ■ ■ ■ ■ ■ ■ ■

.4938

12. PLATE 59. N = 300.
If .4938 of the total number of scores (N) lie between the μ and z-score 2.5, how many scores does this include? _____

■ ■ ■ ■ ■ ■ ■ ■ ■ ■ ■ ■ ■ ■ ■

.4938N = .4938(300) = 148.14

13. Recall that, in a normal distribution, .5 of the scores lie above the μ. PLATE 59. If .4938N lie between the μ and z-score 2.5, the proportion of scores that lies above z-score 2.5 is _____.

■ ■ ■ ■ ■ ■ ■ ■ ■ ■ ■ ■ ■ ■

.0062   (.5000 − .4938)

14. PLATE 59. If .4938N lie between the μ and z-score 2.5, the proportion of N that lies below z-score 2.5 is _____ .

■ ■ ■ ■ ■ ■ ■ ■ ■ ■ ■ ■ ■ ■

.9938   (.5000 + .4938)

---

15. Table 1 presents the proportions of scores between the μ and positive z-scores, which represent proportions of N lying above the μ. Because the normal curve is symmetrical, the proportions are the same for negative z-scores, except that they represent proportions of N lying _____ (above/below) the μ.

■ ■ ■ ■ ■ ■ ■ ■ ■ ■ ■ ■ ■ ■

below

Plate 60.

μ = 50
σ = 10
N = 300

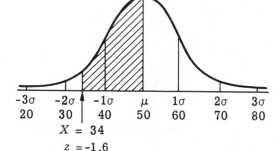

| -3σ | -2σ | -1σ | μ | 1σ | 2σ | 3σ |
| 20 | 30 | 40 | 50 | 60 | 70 | 80 |

$X = 34$
$z = -1.6$

16. For raw score 34, the z-score is −1.6, as determined previously. From Table 1, the proportion of N that lies between the μ and z-score −1.6 is _____ .

■ ■ ■ ■ ■ ■ ■ ■ ■ ■ ■ ■ ■ ■

.4452

126

17. PLATE 60. Table 1 indicates that $.4452N$ lie between the μ and z-score -1.6. This is represented by the shaded area in the plate. Because the z-score is negative, this proportion lies _____ (above/below) the μ.

■ ■ ■ ■ ■ ■ ■ ■ ■ ■ ■ ■ ■ ■

below

18. PLATE 60. You have determined that between score 34 and the μ lie $.4452N$. In this plate, $.4452N =$ _____. This means that between 34 and ____ , there is an N of _____ .

■ ■ ■ ■ ■ ■ ■ ■ ■ ■ ■ ■ ■ ■

133.56, 50, 133.56

A population of 600 students is given an English examination. The μ score for the group is English score 80. The σ of the English scores is 6. How many students received English scores above 89.9? (Assume that the scores are normally distributed.)

19. To determine how many students received English scores above 89.9, you must first determine x for English score 89.9. $x =$ _____ .

■ ■ ■ ■ ■ ■ ■ ■ ■ ■ ■ ■ ■ ■

9.9

20. For English score 89.9, $x = 9.9$. This tells you that the English score 89.9 deviates 9.9 points from the μ. It also tells you that it is _____ (above/below) the μ.

■ ■ ■ ■ ■ ■ ■ ■ ■ ■ ■ ■ ■ ■

above

21. You know that English score 89.9 deviates 9.9 points above the μ because the deviation score (x) is _____ (positive/negative).

■ ■ ■ ■ ■ ■ ■ ■ ■ ■ ■ ■ ■ ■

positive

22. You know that $x = 9.9$. Next, you must determine the z-score which corresponds to the deviation score of 9.9. $z =$ _____ .

■ ■ ■ ■ ■ ■ ■ ■ ■ ■ ■ ■ ■

        1.65   (9.9/6)

23. $x = 9.9$. This yields z-score 1.65. From Table 1, the proportion of students having scores falling between the μ and z-score 1.65 is _____.

■ ■ ■ ■ ■ ■ ■ ■ ■ ■ ■ ■ ■

        .4505

24. $x = 9.9$, z-score 1.65. The proportion of students having scores between the μ and the English score 89.9 is .4505. The proportion of students having English scores *above* English score 89.9 is _____. Recall that $.5N$ lie above the μ.

■ ■ ■ ■ ■ ■ ■ ■ ■ ■ ■ ■ ■

        .0495   (.5000 − .4505)

25. Recall that $N = 600$. The proportion of students having English scores above score 89.9 is .0495. The *number* of students that received scores above 89.9 is _____ .

■ ■ ■ ■ ■ ■ ■ ■ ■ ■ ■ ■

        29.7   (.0495 × 600)

26. As another example, assume a normal distribution in which 7,300 soldiers are each rated on their shooting ability. The μ shooting score for the population is 25, and the σ is 8. The number of soldiers receiving scores below 14.6 is _____. This problem is depicted in Plate 61.

Plate 61.

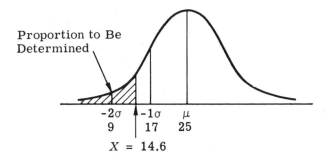

Proportion to Be Determined

-2σ -1σ μ
9   17   25

$X = 14.6$

■ ■ ■ ■ ■ ■ ■ ■ ■ ■ ■ ■ ■

$x = 14.6 - 25 = -10.4$ (Formula 6)

$z = \dfrac{-10.4}{8} = -1.3$ (Formula 10)

For z-score -1.3, $.4032N$ have shooting scores between the μ and 14.6 (Table 1). Since $.5000N$ lie below the μ, $.5000N - .4032N = .0968N$ have shooting scores below 14.6. $N = 7300$. $.0968N = .0968(7300) = 706.64$ soldiers have shooting scores below 14.6.

---

27. To determine how many scores lie above or below a certain score in a normally distributed population, you need to have three measures: the number of scores (N); the mean (μ); and the _____
   _____ (___).

   ■ ■ ■ ■ ■ ■ ■ ■ ■ ■ ■ ■ ■

   standard deviation (σ)

---

28. To say that an individual has a music score of 15 tells you nothing about his music ability. To determine his ability relative to the other musicians in the group, you must know the ____ (symbol), the ____ (symbol), and the ____ (symbol).

   ■ ■ ■ ■ ■ ■ ■ ■ ■ ■ ■ ■ ■

   N, μ, σ (any order)

---

29. George receives the same raw score in two subjects: a music score of 15 and a geometry score of 15. Assume normal distributions for both populations of scores.

$$\text{Music scores} \qquad \mu = 20, \sigma = 5$$
$$\text{Geometry scores} \qquad \mu = 11, \sigma = 2$$

George's music z-score is ___. George's geometry z-score is ___.

■ ■ ■ ■ ■ ■ ■ ■ ■ ■ ■ ■ ■

$$-1, 2$$

---

30. George's z-score in music is -1. The proportion of students receiving music scores *above* George is _____. (Use Table 1.) George's z-score in geometry is 2. The proportion of students receiving geometry scores *above* George is _____.

■ ■ ■ ■ ■ ■ ■ ■ ■ ■ ■ ■ ■

.8413	(.5000 + .3413)
.0228	(.5000 − .4772)

---

31. $.8413N$ received music scores above George. $.0228N$ received geometry scores above George. If $N = 500$, how many students received music scores above George? How many students received geometry scores above George?

■ ■ ■ ■ ■ ■ ■ ■ ■ ■ ■ ■ ■

420.6	($.8413N = .8413 \times 500$)
11.4	($.0228N = .0228 \times 500$)

---

32. In a population of 500 students, 420.6 students received higher music scores than George and 11.4 students received higher geometry scores than George. Conclusion: George is much better in _____ than he is in _____.

■ ■ ■ ■ ■ ■ ■ ■ ■ ■ ■ ■ ■

geometry, music

---

33. Now a method is needed for expressing probability. If you toss a coin a number of times, the chances are that it will turn up heads one-half, or 50%, of the time. That is, the probability of heads occurring is ____%.

■ ■ ■ ■ ■ ■ ■ ■ ■ ■ ■ ■ ■ ■

50

---

34. In statistics, the symbol for probability is P, and the percentage is expressed in decimal form. The probability that a coin will turn up heads may be expressed as $P = .5$, which means that the probability is ____% that it will turn up heads.

■ ■ ■ ■ ■ ■ ■ ■ ■ ■ ■ ■ ■ ■

50

---

35. If the probability of occurrence is 25%, this would be expressed as $P = .25$. If the probability of occurrence is 75%, this would be expressed as $P =$ _____.

■ ■ ■ ■ ■ ■ ■ ■ ■ ■ ■ ■ ■ ■

.75

---

36. If you determined the μ of a population of normally distributed scores, then wrote each score on a slip of paper, placed them in a hat, and then drew one of them out, the probability that the score selected would lie above the μ may be expressed as $P =$ _____ because ____% of the scores lie above the μ in a normal distribution.

■ ■ ■ ■ ■ ■ ■ ■ ■ ■ ■ ■ ■ ■

.50, 50

---

37. The probability of selecting a score above the μ is .5. The probability of selecting a score above Q_1 may be expressed as $P =$ _____ , because ____% of the scores in the distribution lie above Q_1.

■ ■ ■ ■ ■ ■ ■ ■ ■ ■ ■ ■ ■ ■

.75, 75

38. Table 1 can be interpreted as indicating the probability of selecting a score between the μ and the various z-scores. Thus, the probability of selecting a score between the μ and the 1σ may be expressed as $P = .3413$. What is the probability of selecting a score between the μ and 1.6σ? $P =$ _____ .

■ ■ ■ ■ ■ ■ ■ ■ ■ ■ ■ ■ ■ ■

.4452

---

39. TABLE 1. The probability of selecting a score that lies between the μ and 1.6σ is .4452. This means that, out of 100 selections, 44.52 of them would probably have scores that lie between the μ and 1.6σ. How many out of 100 selections would probably have scores between the μ and 1.25σ? _____

■ ■ ■ ■ ■ ■ ■ ■ ■ ■ ■ ■ ■ ■

39.44

---

40. TABLE 1. The probability of selecting a score that lies above 3σ is $P =$ _____ .

■ ■ ■ ■ ■ ■ ■ ■ ■ ■ ■ ■ ■ ■

.0013 $(.5000 - .4987)$

---

41. TABLE 1. The probability of selecting a score that lies below -2.1σ is $P =$ _____ .

■ ■ ■ ■ ■ ■ ■ ■ ■ ■ ■ ■ ■ ■

.0179 $(.5000 - .4821)$

---

42. TABLE 1. The probability of selecting a score that lies between -2σ and 2σ is $P =$ _____ .

■ ■ ■ ■ ■ ■ ■ ■ ■ ■ ■ ■ ■ ■

.9544 $(.4772 + .4772)$

43. The probability of selecting a score that lies between -2σ and 2σ is .9544. How many out of 100 selections would be expected to have scores that lie between these two points? _____

■ ■ ■ ■ ■ ■ ■ ■ ■ ■ ■ ■ ■ ■

95.44

---

44. You are given the following information about a population of scores that is normally distributed: $\mu = 50$, $\sigma = 5$. What is the probability that a score selected at random from this distribution is 57 or more?

To solve this problem, you must first determine the z-score for the raw score of 57.

$$z = \underline{\hspace{1cm}}$$

■ ■ ■ ■ ■ ■ ■ ■ ■ ■ ■ ■ ■ ■

1.4   ($x = 57 - 50; 7/5 = 1.4$)

---

45. For raw score 57, $x = 7$ and $z = 1.4$.

TABLE 1. Between μ and z-score 1.4, $P =$ _____ .

■ ■ ■ ■ ■ ■ ■ ■ ■ ■ ■ ■ ■ ■

.4192

---

46. For raw score 57, $z = 1.4$. Between μ and z-score 1.4, $P = .4192$. You know that above the μ, $P = .5$. Therefore, the probability of obtaining a score of 57 or more is $P =$ _____ .

■ ■ ■ ■ ■ ■ ■ ■ ■ ■ ■ ■ ■ ■

.0808   (.5000 − .4192)

---

47. The probability that a randomly selected score is 57 or more is .0808. This means that if you continued selecting scores at random, _____ % of the time you would select scores that are 57 or more.

■ ■ ■ ■ ■ ■ ■ ■ ■ ■ ■ ■ ■ ■

8.08

---

48. What is the probability that a score selected at random is between 40 and 55? ($\mu = 50$, $\sigma = 5$) To solve this problem you must first determine the z-score for each of the raw scores. For raw score 40, $z = $ ___. For raw score 55, $z = $ ___.

■ ■ ■ ■ ■ ■ ■ ■ ■ ■ ■ ■ ■ ■

-2, 1

---

49. For raw score 40, $z = -2$. For raw score 55, $z = 1$.

TABLE 1. The probability of selecting a score between the μ and the z-score -2 is _____ . The probability of selecting a score between the μ and the z-score 1 is _____ . Therefore, the probability of selecting a score between 40 and 55 is _____ .

■ ■ ■ ■ ■ ■ ■ ■ ■ ■ ■ ■ ■ ■

.4772, .3413, .8185

---

50. The probability that a randomly selected score is between score 40 and 55 is .8185. This means that if you continued selecting scores at random, _____% of the time you would expect to select scores that lie between 40 and 55.

■ ■ ■ ■ ■ ■ ■ ■ ■ ■ ■ ■ ■ ■

81.85

EXERCISES

1. What does the symbol P represent?

2. In a certain arithmetic test there is a probability that one-half of the students will score above 82. How is this probability expressed in decimal form?

3. If there is a 15% chance of getting a grade of A in a certain English course, how can this probability be expressed in decimal form?

4. 95% of the population own television sets. What is the probability, expressed in decimal form, of randomly selecting a person from this population who does not own a television set?

5. Using Table 1, determine the probability of obtaining a score between:
   (a) μ and z-score 1.00
   (b) μ and z-score 2.13
   (c) z-score $-.64$ and z-score 2.05
   (d) z-score 1.75 and z-score 2.25

SAMPLING ERROR

When a sample is drawn from a population, it is unlikely that the statistics derived from the sample are identical to the population parameters. There is always some error involved when a sample is selected from a population. This set will discuss the concept of sampling error, and the new terms, *sampling distribution of means* and *standard error of the mean*, will be introduced.

Interpreting the distribution of sample means allows us to determine probabilities associated with various sample mean values in relation to the population mean. A method for deriving these probabilities will be presented, as well as the relationship of the size of the sample to the standard error of the mean.

SPECIFIC OBJECTIVES OF SET 10

At the conclusion of this set you will be able to:

(1) define the standard error of the mean.

(2) determine from Table 1 the probability of selecting any given sample mean in a sampling distribution.

(3) state the relationship of the number of scores in a sample to the size of the standard error of the mean.

(4) identify the symbol $\sigma_{\bar{X}}$.

1. Suppose, from one population, you select two different samples of equal size. Because it is difficult to obtain a sample that is *exactly* representative of a population, it is likely that the $\overline{X}$'s of these two samples _____ (will/will not) differ.

■ ■ ■ ■ ■ ■ ■ ■ ■ ■ ■ ■ ■

will

2. You may expect that the $\overline{X}$'s of two samples, drawn from the same population, will differ. This is due to *sampling error*, because it is difficult to obtain a sample perfectly _____ of a population.

■ ■ ■ ■ ■ ■ ■ ■ ■ ■ ■ ■ ■

representative

3. The $\overline{X}$'s of samples drawn from the same population are seldom *exactly* the same. This is due to sampling _____ .

■ ■ ■ ■ ■ ■ ■ ■ ■ ■ ■ ■ ■

error

4. If you obtain *many* samples from a population, it is likely that the ___'s (symbol) of the samples will vary.

■ ■ ■ ■ ■ ■ ■ ■ ■ ■ ■ ■ ■

$\overline{X}$

5. This variability among the $\overline{X}$'s of many samples drawn from the same population is due to _____ _____ .

■ ■ ■ ■ ■ ■ ■ ■ ■ ■ ■ ■ ■

sampling error

6. If you draw many samples from a population, the ___'s (symbol) of these samples will _____ due to sampling error.

■ ■ ■ ■ ■ ■ ■ ■ ■ ■ ■ ■ ■

$\overline{X}$, vary

Note: The principle to be illustrated in this set holds true only if all possible samples of a given size are obtained from a population. For ease of presentation, however, Plate 62 presents the $\overline{X}$'s of only twenty-six samples. The assumption should be made that this exhausts the number of possible samples in population.

Plate 62. Means for Twenty-Six Samples, Each Containing N = 50 Scores

4	6	6	3
5	7	4	5
3	4	5	4
1	4	3	3
3	2	4	4
5	4	6	2
5	2		

7. This plate presents the ____'s (symbol) of a large number of _____ selected from the same population.

■ ■ ■ ■ ■ ■ ■ ■ ■ ■ ■ ■ ■

$\overline{X}$, samples

8. PLATE 62. Each number in this plate represents the ____(symbol) of one _____.

■ ■ ■ ■ ■ ■ ■ ■ ■ ■ ■ ■ ■

$\overline{X}$, sample

9. PLATE 62. Each sample whose $\overline{X}$ is presented contains _____ (number) scores.

■ ■ ■ ■ ■ ■ ■ ■ ■ ■ ■ ■ ■

50

A frequency distribution of a set of sample $\overline{X}$'s is called a *sampling distribution of $\overline{X}$'s*. The sampling distribution of the $\overline{X}$'s presented in Plate 62 is given in Plate 63.

Plate 63. Sampling Distribution of $\overline{X}$'s

Frequency distribution of the $\overline{X}$'s of samples selected from the same population. $N = 50$ for each sample.

Sample $\overline{X}$ Values	f
7	1
6	3
5	5
4	8
3	5
2	3
1	1
$\mu = 4$	$\sigma_{\overline{X}} = 1.4$

10. PLATE 63. This frequency distribution is called a _____ distribution of ___'s (symbol) because each value in the distribution is the mean of a sample.

■ ■ ■ ■ ■ ■ ■ ■ ■ ■ ■ ■ ■ ■

sampling, $\overline{X}$

---

11. If you prepare a frequency distribution of all possible sample $\overline{X}$'s, you find that it approximates the shape of a normal curve. Thus, the sampling _____ of $\overline{X}$'s is in the form of a normal distribution.

■ ■ ■ ■ ■ ■ ■ ■ ■ ■ ■ ■ ■ ■

distribution

---

12. PLATE 63. The variability among the $\overline{X}$'s presented in this sampling distribution of $\overline{X}$'s is due to sampling _____ .

■ ■ ■ ■ ■ ■ ■ ■ ■ ■ ■ ■ ■ ■

error

---

13. When you compute the standard deviation of a sampling distribution of $\overline{X}$'s, it is called the *standard error of the $\overline{X}$* because it is a standard measure of the _____ involved in obtaining a sample that is perfectly representative of the population.

■ ■ ■ ■ ■ ■ ■ ■ ■ ■ ■ ■ ■ ■

error

---

14. The symbol for the *standard error of the $\overline{X}$* is written $\sigma_{\overline{X}}$. The symbol σ indicates that the standard error is the standard _____ of a set of sample means.

■ ■ ■ ■ ■ ■ ■ ■ ■ ■ ■ ■ ■ ■

deviation

Note: The use of the Greek symbol $\sigma_{\overline{X}}$ is used here to designate the standard deviation of sample means, computed from the population of sample means. Such sample means would be derived from all possible samples of a given size.

---

15. PLATE 63. The standard error of the mean in this example is $\sigma_{\overline{X}} = $ ____.

■ ■ ■ ■ ■ ■ ■ ■ ■ ■ ■ ■ ■ ■

1.4

---

16. PLATE 63. $\sigma_{\overline{X}} = 1.4$. This indicates that the standard _____ of the $\overline{X}$ for this set of sample $\overline{X}$'s is 1.4.

■ ■ ■ ■ ■ ■ ■ ■ ■ ■ ■ ■ ■ ■

error

---

17. The standard error of the $\overline{X}$ is interpreted in exactly the same manner as the standard _____ of raw scores.

■ ■ ■ ■ ■ ■ ■ ■ ■ ■ ■ ■ ■ ■

deviation

---

18. When you have the $\overline{X}$'s of all possible samples of a given size, the mean of these $\overline{X}$'s will be identical with the population mean (μ). Assuming that all possible samples with $N = 50$ are represented in the sampling distribution of $\overline{X}$'s in Plate 63, the μ of this population is ___.

■ ■ ■ ■ ■ ■ ■ ■ ■ ■ ■ ■ ■ ■

4

---

19. PLATE 63. The standard error of the $\overline{X}$ in this plate is _____ .

■ ■ ■ ■ ■ ■ ■ ■ ■ ■ ■ ■ ■ ■

1.4

---

20. PLATE 63. $\sigma_{\overline{X}} = 1.4$. This indicates that the standard error of the $\overline{X}$ is 1.4 score points. Recall that the standard error of the $\overline{X}$ is interpreted exactly the same as the standard _____ of raw scores.

■ ■ ■ ■ ■ ■ ■ ■ ■ ■ ■ ■ ■ ■

deviation

---

21. Using Table 1, we determined that between the μ and 1σ lie .3413 of the scores. In interpreting the standard error of the $\overline{X}$, the proportion of sample $\overline{X}$'s that lies between the population μ and 1 standard error is also

_____ .

■ ■ ■ ■ ■ ■ ■ ■ ■ ■ ■ ■ ■ ■

.3413

---

22. PLATE 63. The value of the sample $\overline{X}$ that lies one standard error above the μ is _____ .

■ ■ ■ ■ ■ ■ ■ ■ ■ ■ ■ ■ ■ ■

5.4

---

23. PLATE 63. The proportion of sample $\overline{X}$'s that lies between values 4 and 5.4 is _____ .

■ ■ ■ ■ ■ ■ ■ ■ ■ ■ ■ ■ ■ ■

.3413

---

24. Recall that proportions may also be expressed as probabilities. Thus, in Plate 63, if 50%, or .5, of the $\overline{X}$'s lie above the μ, it can be said that the probability of obtaining a sample whose $\overline{X}$ is above the population μ is ___.

■ ■ ■ ■ ■ ■ ■ ■ ■ ■ ■ ■ ■ ■

.5

---

25. The probability of obtaining a $\overline{X}$ that lies above the μ is $P = .5$. The probability of obtaining a $\overline{X}$ that lies between the μ and $1\sigma_{\overline{X}}$ is _____ .

■ ■ ■ ■ ■ ■ ■ ■ ■ ■ ■ ■ ■ ■

.3413

Note: When dealing with μ and $\sigma_{\overline{X}}$, the z-score values are obtained by using $z = (\overline{X} - \mu)/\sigma_{\overline{X}}$.

26. PLATE 63. $\mu = 4, \sigma_{\overline{X}} = 1.4$. The z-score of a value of 6.1 is $z =$ _____ .

■ ■ ■ ■ ■ ■ ■ ■ ■ ■ ■ ■ ■ ■

$$z = \frac{2.1}{1.4} = 1.5$$

---

27. PLATE 63. For value 6.1, $z = 1.5$. Using Table 1, the probability of obtaining a sample $\overline{X}$ that is between the μ and 6.1 is $P =$ _____ .

■ ■ ■ ■ ■ ■ ■ ■ ■ ■ ■ ■ ■ ■

.4332

---

28. PLATE 63. The probability of a sample $\overline{X}$ lying between the μ and 6.1 is $P = .4332$. This means that _____ % of the sample $\overline{X}$'s lie between the μ and 6.1.

■ ■ ■ ■ ■ ■ ■ ■ ■ ■ ■ ■ ■ ■

43.32

29. PLATE 63. To determine the probability of obtaining a sample $\overline{X}$ that is between score value 2.1 and μ, first compute the z score. $z =$ _____ . $P =$ _____ .

■ ■ ■ ■ ■ ■ ■ ■ ■ ■ ■ ■ ■ ■

$\qquad$ −1.36 $\qquad\qquad\qquad$ $(x = 2.1 − 4 = −1.9)$
$\qquad$ .4131 $\qquad\qquad\qquad$ $(z = −1.9/1.4 = −1.36)$

30. The probability of obtaining a sample $\overline{X}$ that is between 2.1 and μ is $P = .4131$. The probability of obtaining a $\overline{X}$ that lies between the μ and 6.1 is $P = .4332$. Therefore, the probability of obtaining a $\overline{X}$ that lies between 2.1 and 6.1 is _____ .

■ ■ ■ ■ ■ ■ ■ ■ ■ ■ ■ ■ ■ ■

.8463

31. PLATE 63. The probability of obtaining a sample whose $\overline{X}$ lies between 2.6 and 6.8 is $P =$ _____ .

■ ■ ■ ■ ■ ■ ■ ■ ■ ■ ■ ■ ■ ■

From Table 1

$$2.6 − 4 = −1.4 \qquad z = \frac{−1.4}{1.4} = −1 \qquad P = .3413$$

$$6.8 − 4 = 2.8 \qquad z = \frac{2.8}{1.4} = 2 \qquad P = .4772$$

$$P = .3413 + .4772 = .8185$$

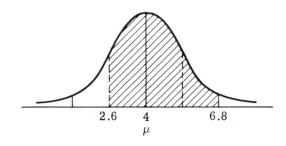

32. In a sampling distribution of $\overline{X}$'s in which the μ is 4 and the $\sigma_{\overline{X}}$ is 1.4, the probability of obtaining a sample whose $\overline{X}$ lies between 2.6 and 6.8 is

_____ .

■ ■ ■ ■ ■ ■ ■ ■ ■ ■ ■ ■ ■

.8185

33. If you know the μ and the _____ _____ of the $\overline{X}$ you can determine the probability of a sample mean lying between any two selected score values.

■ ■ ■ ■ ■ ■ ■ ■ ■ ■ ■ ■ ■

standard error

34. PLATE 63. Each of the twenty-six samples whose means are presented contains _____ raw scores.

■ ■ ■ ■ ■ ■ ■ ■ ■ ■ ■ ■ ■

50

35. If the number of scores within each of the samples is increased, it logically follows that there is _____ (more/less) error involved in the sampling.

■ ■ ■ ■ ■ ■ ■ ■ ■ ■ ■ ■ ■

less

36. There is less error involved in a large sample than in a small sample. Therefore, in a sampling distribution of $\overline{X}$'s with each sample having thirty scores, the $\sigma_{\overline{X}}$ will be _____ (less/greater) than where each sample contains fifty scores.

■ ■ ■ ■ ■ ■ ■ ■ ■ ■ ■ ■ ■

greater

Plate 64 presents three sampling distributions of $\overline{X}$'s.

Plate 64. Sampling Distribution of $\overline{X}$'s

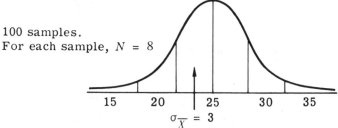

100 samples.
For each sample, $N = 8$

15   20   25   30   35

$\sigma_{\overline{X}} = 3$

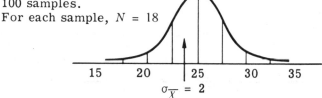

100 samples.
For each sample, $N = 18$

15   20   25   30   35

$\sigma_{\overline{X}} = 2$

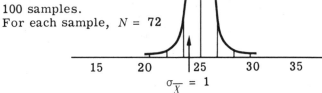

100 samples.
For each sample, $N = 72$

15   20   25   30   35

$\sigma_{\overline{X}} = 1$

37. PLATE 64. In the top sampling distribution, where each sample contains eight scores, the $\sigma_{\overline{X}} = 3$; whereas, in the second distribution, where each sample contains eighteen scores, the $\sigma_{\overline{X}} = $ ___.

■ ■ ■ ■ ■ ■ ■ ■ ■ ■ ■ ■ ■ ■

2

---

38. PLATE 64. In the bottom sampling distribution of $\overline{X}$'s where each sample contains seventy-two scores, the $\sigma_{\overline{X}} = $ ___.

■ ■ ■ ■ ■ ■ ■ ■ ■ ■ ■ ■ ■ ■

1

---

39. PLATE 64. From these three sampling distributions of $\overline{X}$'s, it can be seen that as the sample size is increased, the $\sigma_{\overline{X}}$ is _____.

■ ■ ■ ■ ■ ■ ■ ■ ■ ■ ■ ■ ■ ■

decreased

145

40. PLATE 64. As the number of scores in each sample is increased, there is _____ (more/less) error in the sample $\overline{X}$'s.

■ ■ ■ ■ ■ ■ ■ ■ ■ ■ ■ ■ ■ ■

less

---

41. PLATE 64. Here it can be seen that the larger the sample size, the _____ (more/less) confidence you have that any particular sample $\overline{X}$ approximates the μ.

■ ■ ■ ■ ■ ■ ■ ■ ■ ■ ■ ■ ■ ■

more

EXERCISES

1. What is meant by the term *standard error of the mean*? What is its symbol?

2. Using Table 1, determine the probability of selecting a sample whose mean score lies
   (a) between μ and $2\sigma_{\overline{X}}$
   (b) between $-1\sigma_{\overline{X}}$ and $2\sigma_{\overline{X}}$
   (c) between $-2.05\sigma_{\overline{X}}$ and $.5\sigma_{\overline{X}}$

3. Assume the μ of a set of data is 12 and the $\sigma_{\overline{X}}$ is 3. Using Table 1, determine the probability that you will select a sample whose mean score lies
   (a) between 9 and 12
   (b) between 12 and 16
   (c) between 10 and 13

4. If the size of a sample is increased, how is the standard error of the mean affected?

ESTIMATION OF THE POPULATION VARIANCE

The last set presented a method for determining the probabilities associated with any given sample mean when the population mean is known. In research, however, we are almost never fortunate enough to know the population mean. We usually have only the data contained in one sample from which to infer values of the population parameters; and before we can estimate the population mean, we must estimate the variability in the population.

This set presents the method for estimating the variance of a population when you have data for only one sample of that population. There is an introduction to the concept of *sum of squares* and its calculation by the raw score and deviation score methods. We will also distinguish between the *biased* estimate and the *unbiased* estimate of the population variance based on a single sample, and explain how the concept of *degrees of freedom* is used to derive unbiased estimates of the population variance.

SPECIFIC OBJECTIVES OF SET 11

At the conclusion of this set you will be able to:

(1)   use Formula 11a to calculate the sum of squares for a frequency distribution by the deviation score method.

(2)   use Formula 11b to calculate the sum of squares for a frequency distribution by the raw score method.

(3)   determine the degrees of freedom for a frequency distribution.

(4) use Formula 12a to calculate an unbiased estimate of population variance from a sample frequency distribution by the deviation score method.

(5) use Formula 12b to calculate an unbiased estimate of the population variance from a sample frequency distribution by the raw score method.

(6) identify the symbols s^2, σ^2, and df.

1. In the examples that have been presented, the standard error of the $\overline{X}$ has been obtained by computing the standard_____of all possible sample $\overline{X}$'s.

■ ■ ■ ■ ■ ■ ■ ■ ■ ■ ■ ■ ■

deviation

2. You seldom have a large number of samples; however, from the data contained in one sample you can estimate the value of the population _____ _____ (parameters/statistics).

■ ■ ■ ■ ■ ■ ■ ■ ■ ■ ■ ■ ■

parameters

3. In order to make estimates of population _____ (parameters/statistics) from the _____ (parameters/statistics) of one sample, you need to learn some new terms. The first term is *sum of squares.*

■ ■ ■ ■ ■ ■ ■ ■ ■ ■ ■ ■ ■

parameters, statistics

Formula 11. Calculation of the Sum of Squared Deviations ("Sum of Squares") (Formulas 11a and 11b are Equivalent)

Deviation Score Method $\Sigma x^2 = \Sigma (X - \overline{X})^2$ (Formula 11a)

Raw Score Method $\Sigma x^2 = \Sigma X^2 - \dfrac{(\Sigma X)^2}{N}$ (Formula 11b)

---

4. FORMULA 11. These formulas for the "sum of squares" use Σx^2 instead of $\Sigma f x^2$, as used in Formula 8a. The f was included earlier only for illustrating the computation. The two expressions Σx^2 and $\Sigma f x^2$ mean exactly the same thing: the sum of the squared _____ scores.

■ ■ ■ ■ ■ ■ ■ ■ ■ ■ ■ ■ ■

deviation

5. FORMULA 11. These two formulas give two methods of calculating Σx^2: the deviation score method and the _____ _____ method.

■ ■ ■ ■ ■ ■ ■ ■ ■ ■ ■ ■ ■

raw score

6. FORMULA 11. These two formulas are equivalent and yield identical values. The raw score formula is given because it is sometimes easier to use in computation. The expression for "sum of squares" is _____ (symbols).

■ ■ ■ ■ ■ ■ ■ ■ ■ ■ ■ ■ ■

$$\Sigma x^2$$

---

7. FORMULA 11. The expression, Σx^2, is called the "sum of squares" and indicates that you _____ the squared deviation scores of the sample.

■ ■ ■ ■ ■ ■ ■ ■ ■ ■ ■ ■ ■

sum

---

8. FORMULA 11. Σx^2 is called the _____ of _____ .

■ ■ ■ ■ ■ ■ ■ ■ ■ ■ ■ ■ ■

sum, squares

Formula 12. Estimate of the Population Variance from Sample Data (Formulas 12a and 12b are Equivalent)

Deviation Score Method $\quad s^2 = \dfrac{\Sigma x^2}{N-1}$ $\qquad$ (Formula 12a)

Raw Score Method $\qquad s^2 = \dfrac{N\Sigma X^2 - (\Sigma X)^2}{N(N-1)}$ $\quad$ (Formula 12b)

9. Recall that the symbol for the sample standard deviation is s'. The symbol for the sample variance is s'^2. Formula 12 presents the formula for *estimating* the population variance from sample data. The symbol used to represent this variance estimate is _____ .

■ ■ ■ ■ ■ ■ ■ ■ ■ ■ ■ ■ ■

$$s^2$$

---

10. The symbol for the population variance is σ^2. The symbol for the *estimate* of the population variance, from sample data, is _____ (symbol).

■ ■ ■ ■ ■ ■ ■ ■ ■ ■ ■ ■ ■

$$s^2$$

---

11. The formula for the variance of a population is:

$$\sigma^2 = \frac{\Sigma x^2}{N}$$

The formula for the estimate of the population variance, from sample data, as presented in Formula 12a, is:

$$s^2 = \frac{\overline{\quad\quad}}{\underline{\quad\quad}}$$

■ ■ ■ ■ ■ ■ ■ ■ ■ ■ ■ ■ ■ ■

$$\frac{\Sigma x^2}{N-1}$$

12.

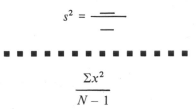

$$\sigma^2 = \frac{\Sigma x^2}{N}$$

Formula 12a, for estimating the population variance from sample data, differs from the above formula for the population variance, in that it uses $N - 1$ as the denominator term instead of ____(symbol).

■ ■ ■ ■ ■ ■ ■ ■ ■ ■ ■ ■ ■ ■

N

13. FORMULA 12a. If you use N instead of $N - 1$ in the formula, it would tend to be an underestimate of the population variance. Hence, it would be called a *biased* _____ of the population variance.

■ ■ ■ ■ ■ ■ ■ ■ ■ ■ ■ ■ ■ ■

estimate

14. The use of $N - 1$ in the denominator of Formula 12a yields an *unbiased* estimate of s^2. Thus, Formula 12a yields a _____ (more/less) accurate estimate of the population variance.

■ ■ ■ ■ ■ ■ ■ ■ ■ ■ ■ ■ ■ ■

more

15. FORMULA 12a. The term, $N - 1$, indicates that you subtract ___ from the number of scores in the sample. A discussion of the meaning of this term follows.

■ ■ ■ ■ ■ ■ ■ ■ ■ ■ ■ ■ ■ ■ ■

1

Plate 65.

Example A:	10	20	30	40	50	$\bar{X} = 30$
Example B:	16	22	35	37	(40)	$\bar{X} = 30$
Example C:	5	11	37	45	○	$\bar{X} = 30$

16. PLATE 65. To illustrate the concept of degrees of freedom, first compute the $\bar{X}$ of the five scores in Example A. $\bar{X} =$ ___.

■ ■ ■ ■ ■ ■ ■ ■ ■ ■ ■ ■ ■ ■ ■

30

---

17. PLATE 65. EXAMPLE A. Now compute the deviation of each score from the $\bar{X}$. The five deviation scores are ___ , ___ , ___ , ___ , and ___ .

■ ■ ■ ■ ■ ■ ■ ■ ■ ■ ■ ■ ■ ■ ■

−20, −10, 0, 10, and 20

---

18. PLATE 65. EXAMPLE A. The five deviation scores are −20, −10, 0, 10, and 20. It is evident that the sum of the deviation scores is ___.

■ ■ ■ ■ ■ ■ ■ ■ ■ ■ ■ ■ ■ ■ ■

0

---

19. PLATE 65. EXAMPLE A. If the first four measurements are 10, 20, 30, and 40, the fifth can have only one value in order for the sum of the deviations to equal zero. That value is ___.

■ ■ ■ ■ ■ ■ ■ ■ ■ ■ ■ ■ ■ ■ ■

50

---

20. PLATE 65. EXAMPLE A. If $\bar{X} = 30$, four of the scores may be of any value, but the fifth score is not _____ to vary.

■ ■ ■ ■ ■ ■ ■ ■ ■ ■ ■ ■ ■ ■

free

---

21. PLATE 65. EXAMPLE B. If $\bar{X} = 30$ and four of the scores have values 16, 22, 35, and 37, then in order for the sum of the deviation scores to equal zero the fifth score must have the value ____.

■ ■ ■ ■ ■ ■ ■ ■ ■ ■ ■ ■ ■ ■

40

---

22. PLATE 65. EXAMPLE C. In order for the sum of the deviation scores to equal zero, the fifth score in this sample must be ____ .

■ ■ ■ ■ ■ ■ ■ ■ ■ ■ ■ ■ ■ ■

52

---

23. Thus, for any set of data we may generalize that the sum of the _____ scores must equal zero.

■ ■ ■ ■ ■ ■ ■ ■ ■ ■ ■ ■ ■ ■

deviation

---

24. As illustrated in the above examples, if there are five scores in a sample, four of these scores may be of any value, but the fifth is not _____ to vary.

■ ■ ■ ■ ■ ■ ■ ■ ■ ■ ■ ■ ■ ■

free

---

25. FORMULA 12a. This formula indicates that to obtain s^2 you divide Σx^2, which is called the _____ of _____, by $N - 1$, which is called the _____ of _____ .

■ ■ ■ ■ ■ ■ ■ ■ ■ ■ ■ ■ ■ ■

sum, squares, degrees, freedom

26. When you divide Σx^2 by $N - 1$, you obtain an unbiased estimate of the _____ of the scores in the population.

■ ■ ■ ■ ■ ■ ■ ■ ■ ■ ■ ■ ■ ■

variance

---

27. When you *know* the variance of a population, you use ____ (symbol). When you *estimate* the variance of a population from data contained in a sample, you use ____ (symbol).

■ ■ ■ ■ ■ ■ ■ ■ ■ ■ ■ ■ ■ ■

σ^2 , s^2

---

28. If you have data from a sample and wish to make an estimate of the variance of the population from which the sample was selected, you use Formula 12. The symbol s^2 is used in this formula to indicate that the variance is estimated from data contained in a _____ (sample/population).

■ ■ ■ ■ ■ ■ ■ ■ ■ ■ ■ ■ ■ ■

sample

---

29. Using Formula 12a, compute the variance estimate from the following sample data: $N = 81$, $\Sigma x^2 = 120$.

$$s^2 = \underline{\hspace{4cm}}$$

■ ■ ■ ■ ■ ■ ■ ■ ■ ■ ■ ■ ■ ■

$$\frac{120}{81 - 1} = 1.5$$

30. Often it is cumbersome, particularly for large samples, to use the deviation method to obtain s^2. Formula 12b presents an algebraically equivalent raw score formula for computing s^2. Using Formula 12b, compute the variance estimate from the following sample data: $N = 81$, $\Sigma X = 197$, $\Sigma X^2 = 599$.

$$s^2 = \underline{\hspace{4cm}}$$

■ ■ ■ ■ ■ ■ ■ ■ ■ ■ ■ ■ ■

$$\frac{81(599) - (197)^2}{80(81)} = 1.5$$

1. For the following frequency distributions, calculate the "sum of squares" by the deviation score method, using Formula 11a, and by the raw score method, using Formula 11b. Check to make certain that your answers are identical for the two methods.

(a) X	f	(b) interval	f
10	2	26-30	2
9	3	21-25	3
8	1	16-20	5
7	4	11-15	4
6	2	6-10	3
5	2	1-5	3
4	1		
3	1		

2. How many degrees of freedom are associated with the frequency distributions in Exercises 1(a) and 1(b)?

3. Calculate the unbiased estimate of the population variance for the distributions in Exercises 1(a) and 1(b). Use Formula 12a.

4. Calculate the unbiased estimate of the population variance for the distributions in Exercises 1(a) and 1(b). Use the raw score method as presented in Formula 12b. Check your answers with those obtained in Exercise 3.

5. What do the symbols s^2, σ^2, and df represent?

ESTIMATION OF THE STANDARD ERROR OF THE MEAN / 95% CONFIDENCE INTERVAL

It has been pointed out earlier that, if there are many samples, the standard deviation of the sample means can be calculated directly to obtain the standard error of the mean. This, of course, is highly impractical: usually we are unable to obtain the values of population parameters. This set will demonstrate that, from the data in one sample, we can *estimate* the standard error of the mean. Using this estimate we can construct a "confidence interval" which will permit us to make certain statements regarding the value of the population mean.

The method of determining the 95% confidence interval from sample data is presented, as well as its interpretation relative to the population mean.

SPECIFIC OBJECTIVES OF SET 12

At the conclusion of this set you will be able to:

(1) use Formula 13 to derive an estimate of the standard error of the mean from the data in one sample.

(2) state the generally accepted ‚cut-off level of probability for accepting a hypothesis.

(3) calculate the 95% confidence interval from the data in one sample.

(4) identify the symbol $s_{\overline{X}}$.

1. In the earlier discussion, the standard error of the mean ($\sigma_{\overline{X}}$) was obtained by taking the $\overline{X}$'s of all possible samples of a certain size and computing their _____ deviation.

■ ■ ■ ■ ■ ■ ■ ■ ■ ■ ■ ■ ■ ■

standard

Formula 13. Estimate of the Standard Error of the Mean

$$s_{\overline{X}} = \frac{s}{\sqrt{N}}$$

---

2. Generally, you do not have a large number of $\overline{X}$'s to use in the computation of $\sigma_{\overline{X}}$. However, with the data in one sample, you can use Formula 13 to estimate the standard error of the _____ .

■ ■ ■ ■ ■ ■ ■ ■ ■ ■ ■ ■ ■ ■

mean

---

3. FORMULA 13. The symbol for the estimate of the standard error of the mean, from sample data, is _____ (symbol).

■ ■ ■ ■ ■ ■ ■ ■ ■ ■ ■ ■ ■ ■

$s_{\overline{X}}$

---

4. FORMULA 13. The symbol, $s_{\overline{X}}$, is used in this formula because it is an estimate of the standard error of the mean, derived from _____ (sample/population) data.

■ ■ ■ ■ ■ ■ ■ ■ ■ ■ ■ ■ ■ ■

sample

---

5. The $s_{\overline{X}}$ is a _____ (statistic/parameter).

■ ■ ■ ■ ■ ■ ■ ■ ■ ■ ■ ■ ■ ■ ■

statistic

---

6. The $s_{\bar{X}}$ is a statistic because it is derived from the data of a _____ (sample/population).

■ ■ ■ ■ ■ ■ ■ ■ ■ ■ ■ ■ ■ ■

sample

---

7. In Formula 13, the symbol s represents an estimate of the population standard deviation of _____ scores; this estimate has been derived from the data in one sample.

■ ■ ■ ■ ■ ■ ■ ■ ■ ■ ■ ■ ■ ■

raw

---

8. In Formula 13, the symbol N represents the _____ of raw scores in the sample.

■ ■ ■ ■ ■ ■ ■ ■ ■ ■ ■ ■ ■ ■

number

---

9. In Formula 13, the two statistics of a sample that you need in order to compute the $s_{\bar{X}}$ are the ___ (symbol) and the ___ (symbol).

■ ■ ■ ■ ■ ■ ■ ■ ■ ■ ■ ■ ■ ■

s, N

---

10. Formula 13 indicates that the $s_{\bar{X}}$ is calculated by dividing ___ (symbol) by the square root of ___ (symbol).

■ ■ ■ ■ ■ ■ ■ ■ ■ ■ ■ ■ ■ ■

s, N

Note: Whenever you need the square root of a number, consult Table 7 at the rear of the text.

11. Consider a sample in which $N = 100$, $\overline{X} = 12$, and $s = 16$. Set up the computation for $s_{\overline{X}}$ by substituting the numerical values for the symbols in Formula 13.

$$s_{\overline{X}} = \frac{s}{\sqrt{N}} = \frac{\overline{}}{\sqrt{}}$$

■ ■ ■ ■ ■ ■ ■ ■ ■ ■ ■ ■ ■

$$\frac{16}{\sqrt{100}}$$

12. Sample: $N = 100$, $\overline{X} = 12$, $s = 16$.

FORMULA 13. Do the necessary computation to obtain the $s_{\overline{X}}$.

$$s_{\overline{X}} = \frac{16}{\sqrt{100}} = \underline{}$$

■ ■ ■ ■ ■ ■ ■ ■ ■ ■ ■ ■ ■

1.6

13. Sample: $N = 100$, $\overline{X} = 12$, $s = 16$.

The standard error of the mean $s_{\overline{X}}$ is 1.6. This means that the estimate of the standard deviation of a population of sample $\overline{X}$'s, with each $N = 100$, is

_____ .

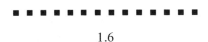

1.6

14. Sample: $N = 100$, $\overline{X} = 12$, $s = 16$, $s_{\overline{X}} = 1.6$.

If the sample from which these statistics are derived is representative of the population, we can infer population _____ from them.

■ ■ ■ ■ ■ ■ ■ ■ ■ ■ ■ ■ ■ ■

parameters

---

15. When you use statistics to make an inference regarding a population _____ (parameter/statistic), from a sample_____ (parameter/statistic), it is called a *statistical inference.*

■ ■ ■ ■ ■ ■ ■ ■ ■ ■ ■ ■ ■ ■

parameter, statistic

---

16. The process of using a sample statistic for making inferences regarding a population parameter is called statistical _____ .

■ ■ ■ ■ ■ ■ ■ ■ ■ ■ ■ ■ ■ ■

inference

---

17. You know that a sample generally is not *perfectly* representative of a population. This is due to _____ error.

■ ■ ■ ■ ■ ■ ■ ■ ■ ■ ■ ■ ■ ■

sampling

---

18. The statistic that you have just learned which gives you a measure of the error involved in sampling is called the _____ _____ of the ___ (symbol).

■ ■ ■ ■ ■ ■ ■ ■ ■ ■ ■ ■ ■ ■

standard error, $\overline{X}$

---

19. In cases where $s_{\overline{X}}$ is computed from a very large sample, the probability of obtaining a sample $\overline{X}$ between the μ and any value of $s_{\overline{X}}$ can be determined from the properties of the normal curve as presented in Table 1. Thus, if $s_{\overline{X}}$ is derived from a large sample, the probability of obtaining a sample $\overline{X}$ between μ and $1s_{\overline{X}}$ is _____ .

■ ■ ■ ■ ■ ■ ■ ■ ■ ■ ■ ■ ■ ■

.3413

20. On the other hand, suppose we wish to make a statement regarding the value of μ when it is *not* known. One way to make such an estimate is to take a random sample, compute $\overline{X}$, and state that this $\overline{X}$ is the best _____ of the value of μ.

■ ■ ■ ■ ■ ■ ■ ■ ■ ■ ■ ■ ■ ■

estimate

21. We know, however, that we can have little confidence that the obtained $\overline{X}$ is exactly the value of the unknown μ, due to _____ _____ .

■ ■ ■ ■ ■ ■ ■ ■ ■ ■ ■ ■ ■ ■

sampling error

22. It would seem that we could determine $\overline{X}$ and $s_{\overline{X}}$ and, using Table 1, specify an *interval* within which we could have a degree of confidence that the ___ (symbol) is likely to be found.

■ ■ ■ ■ ■ ■ ■ ■ ■ ■ ■ ■ ■ ■

μ

23. Recall the situation in which we *knew* the μ and $\sigma_{\overline{X}}$ and could state the probability of obtaining a sample $\overline{X}$ within any specified interval under the normal curve. In contrast, if we obtain a sample $\overline{X}$ we cannot legitimately speak of the probability that the μ lies within any specified interval because of the fact that, for a given population, μ is _____ (one/many) value(s) and has ____ (a/no) distribution.

■ ■ ■ ■ ■ ■ ■ ■ ■ ■ ■ ■ ■ ■

one, no

---

24. Therefore, because μ ___(is/is not) a specific value, although unknown, we _____(can/cannot) make a statement regarding the probability that it will fall within a certain interval.

■ ■ ■ ■ ■ ■ ■ ■ ■ ■ ■ ■ ■

is, cannot

---

25. However, we can determine what is termed a *confidence interval*, using sample statistics, and make certain statements regarding the μ. First, we will illustrate the method of determining a confidence interval and then will interpret its meaning. We shall determine the 95% confidence interval, given the data in one sample, as follows: $N = 100$, $\overline{X} = 10$, $s_{\overline{X}} = 2$. The 95% confidence interval is depicted in Plate 66.

Plate 66.

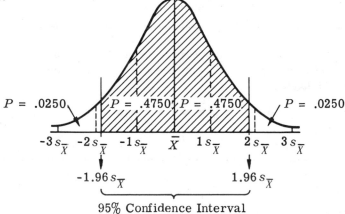

In contrast with previous curves, notice that the mean of this distribution curve is designated as ___ ($\mu/\overline{X}$).

■ ■ ■ ■ ■ ■ ■ ■ ■ ■ ■ ■ ■

$\overline{X}$

26. The shaded area under the normal curve in Plate 66 gives the area defining the 95% confidence interval. The probability that a value in this distribution lies outside this interval is .05. The probability that a value lies within it is _____ .

■ ■ ■ ■ ■ ■ ■ ■ ■ ■ ■ ■ ■ ■

.95

---

27. PLATE 66. For the 95% confidence interval, the probability that a value lies above the $\overline{X}$ is $P =$ _____ , and the probability that a value lies below the $\overline{X}$ is $P =$ _____ .

■ ■ ■ ■ ■ ■ ■ ■ ■ ■ ■ ■ ■ ■

.4750, .4750

---

28. PLATE 66. For the 95% confidence interval, the probability that a value lies above the $\overline{X}$ is .4750. Checking the "μ to z" column of Table 1 for $P = .4750$, such a value can lie _____ $s_{\overline{X}}$ from the $\overline{X}$.

■ ■ ■ ■ ■ ■ ■ ■ ■ ■ ■ ■ ■ ■

1.96

---

29. PLATE 66. For the 95% confidence interval, the probability that a value lies below the $\overline{X}$ is also .4750, which means it can also lie _____ $s_{\overline{X}}$ from the $\overline{X}$.

■ ■ ■ ■ ■ ■ ■ ■ ■ ■ ■ ■ ■ ■

−1.96

---

30. PLATE 66. Thus, for the 95% confidence interval, a value can lie from _____ $s_{\overline{X}}$ to _____ $s_{\overline{X}}$.

■ ■ ■ ■ ■ ■ ■ ■ ■ ■ ■ ■ ■ ■

−1.96, 1.96

31. PLATE 66. The 95% confidence interval is from $-1.96\ s_{\bar{X}}$ to $1.96\ s_{\bar{X}}$. This interval is usually written ±1.96, which indicates that the interval is from $-1.96\ s_{\bar{X}}$ to + _____ _____ (symbol).

■ ■ ■ ■ ■ ■ ■ ■ ■ ■ ■ ■ ■ ■

$1.96\ s_{\bar{X}}$

---

32. PLATE 66. The 95% confidence interval is represented by the two values which lie ± _____ __ (symbol) from the $\bar{X}$.

■ ■ ■ ■ ■ ■ ■ ■ ■ ■ ■ ■ ■ ■

$1.96\ s_{\bar{X}}$

---

33. Example: $\bar{X}$ = 10, $s_{\bar{X}}$ = 2. Because $1 s_{\bar{X}} = 2$ score points, then $1.96 s_{\bar{X}}$ = _____ score points.

■ ■ ■ ■ ■ ■ ■ ■ ■ ■ ■ ■ ■ ■

3.92

---

34. Example: $\bar{X}$ = 10, $s_{\bar{X}}$ = 2. The 95% confidence interval is $\pm 1.96\ s_{\bar{X}}$. $\pm 1.96\ s_{\bar{X}}$ lies 3.92 points above and below the $\bar{X}$. These two values are _____ and _____ .

■ ■ ■ ■ ■ ■ ■ ■ ■ ■ ■ ■ ■ ■

13.92, 6.08

---

35. Now, if we select another sample of the same size from the same population and compute $\bar{X}$ and $s_{\bar{X}}$, we _____ (are/are not) likely to obtain exactly the same sample statistics.

■ ■ ■ ■ ■ ■ ■ ■ ■ ■ ■ ■ ■ ■

are not

---

36. Because a second sample will not yield identical $\bar{X}$ and $s_{\bar{X}}$ to those of the first sample, the 95% confidence interval computed from the data in the second sample will be _____ (different/the same).

■ ■ ■ ■ ■ ■ ■ ■ ■ ■ ■ ■ ■ ■

different

---

37. We could continue to take samples of the same size and determine the 95% confidence interval from the data in each of them. We would then have a collection of confidence intervals, each derived from a _____. Some of these intervals will contain the ___(symbol) and some will not.

■ ■ ■ ■ ■ ■ ■ ■ ■ ■ ■ ■ ■ ■

sample, μ

---

38. In other words, we can state that ____ % of the intervals obtained by this method will contain the value of ___(symbol).

■ ■ ■ ■ ■ ■ ■ ■ ■ ■ ■ ■ ■ ■

95, μ

---

39. It is important to be clear regarding the interpretation of the meaning of a confidence interval. It is *not* a statement about the probability that μ will fall within the given interval. Rather, it is a probability statement about the percentage of similarly derived _____ which contain the value of μ.

■ ■ ■ ■ ■ ■ ■ ■ ■ ■ ■ ■ ■ ■

intervals

---

40. This is because there are as many possible intervals as there are possible _____. On the other hand, there is but one population with a unique ___(symbol).

■ ■ ■ ■ ■ ■ ■ ■ ■ ■ ■ ■ ■ ■

samples, μ

EXERCISES

1. Using Formula 13, calculate the standard error of the mean from the following sets of sample data.

    Sample (a) $N = 81$ $s = 4.5$

    Sample (b) $N = 144$ $s = 21$

2. In Exercise 1(a) suppose that $\bar{X} = 15$. Calculate the 95% confidence interval for this distribution.

3. In Exercise 1(b) suppose that $\bar{X} = 200$. Calculate the 95% confidence interval for this distribution.

4. What is the generally accepted cut-off level of probability for accepting a hypothesis?

5. What does the symbol $s_{\bar{X}}$ represent?

99% CONFIDENCE INTERVAL/
t DISTRIBUTION

The 95% confidence interval makes it possible for us to make certain statements regarding the value of the population mean. A more stringent confidence interval, with an accompanying higher probability, is the 99% confidence interval given in this set.

The calculation of the 95% and 99% confidence limits, using Table 1, have been solely for use with large samples. When the number of raw scores in a sample is few (approximately thirty or less), the probability values associated with the normal distribution are inappropriate for establishing confidence intervals. For each sample size there is a unique distribution called the *t* distribution. The nature of the *t* distribution is discussed in this set, as well as a method for determining confidence intervals using data derived from small samples.

SPECIFIC OBJECTIVES OF SET 13

At the conclusion of this set you will be able to:

(1) state the two confidence intervals generally employed by researchers.

(2) calculate the 99% confidence interval from the data in one sample with a large *N*.

(3) calculate from Table 2 the 95% confidence interval for a sample with a small *N*.

(4) calculate from Table 2 the 99% confidence interval for a sample with a small N.

(5) describe the nature of the t distribution and its relationship to sample size.

1. The 95% confidence interval is represented by $\pm 1.96 \, s_{\bar{X}}$, where $s_{\bar{X}}$ is computed from a large sample. This is interpreted as the interval between a value that lies_____ ___ (term) from the $\bar{X}$ and a value that lies _____ ___ (term) from the $\bar{X}$.

■ ■ ■ ■ ■ ■ ■ ■ ■ ■ ■ ■ ■

$-1.96 \, s_{\bar{X}}, +1.96 \, s_{\bar{X}}$

2. The 95% confidence interval is that interval within which the probability that a value lies is $P =$ _____ .

■ ■ ■ ■ ■ ■ ■ ■ ■ ■ ■ ■ ■

.95

3. If you say that the 95% confidence interval for a set of data contains values 5 to 15, the probability that a value lies between 5 and 15 is $P =$ _____ .

■ ■ ■ ■ ■ ■ ■ ■ ■ ■ ■ ■ ■

.95

4. The probability that a value lies *below* 5 or *above* 15 is $P =$ _____ .

■ ■ ■ ■ ■ ■ ■ ■ ■ ■ ■ ■ ■

.05

5. Thus, the probability of randomly selecting a value that is not contained within the 95% confidence interval is _____ .

■ ■ ■ ■ ■ ■ ■ ■ ■ ■ ■ ■ ■

.05

6. A more stringent confidence interval that is commonly used is the 99% confidence interval. When you determine the 99% confidence interval, you know that the probability of a value lying within the interval is $P =$ _____ . The probability that a value lies above or below the interval is $P =$ _____ .

■ ■ ■ ■ ■ ■ ■ ■ ■ ■ ■ ■ ■ ■

.99, .01

7. For the 99% confidence interval, the probability that a value lies within the interval is $P = .99$. Therefore, the probability that a value lies within the confidence interval and above the $\overline{X}$ is $P =$ _____ , and the probability that a value lies within the confidence interval and below the $\overline{X}$ is _____ .

■ ■ ■ ■ ■ ■ ■ ■ ■ ■ ■ ■ ■ ■

.495, .495

8. From Table 1, in a normal distribution the z-score associated with a probability of .495 is _____ .

■ ■ ■ ■ ■ ■ ■ ■ ■ ■ ■ ■ ■ ■

2.58 (for $P = .4951$)

9. Therefore, in a normal distribution the 99% confidence interval lies ± _____ $s_{\overline{X}}$.

■ ■ ■ ■ ■ ■ ■ ■ ■ ■ ■ ■ ■ ■

2.58

10. The 99% confidence interval is represented by ±2.58 $s_{\overline{X}}$ in cases where $s_{\overline{X}}$ is computed from a large sample. There is a probability of $P =$ _____ that any given value lies within the 99% confidence interval.

■ ■ ■ ■ ■ ■ ■ ■ ■ ■ ■ ■ ■ ■

.99

Plate 67.

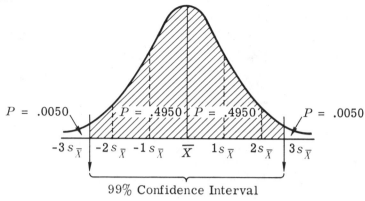

99% Confidence Interval

---

11. The shaded area of this plate indicates the ____ % _____
    _____ .

    ■ ■ ■ ■ ■ ■ ■ ■ ■ ■ ■ ■ ■ ■ ■

    99% confidence interval

---

12. For the 99% confidence interval, the probability that a value lies below
    $-2.58\ s_{\overline{X}}$ is $P =$ _____, and the probability that it lies above $2.58\ s_{\overline{X}}$ is
    $P =$ _____ .

    ■ ■ ■ ■ ■ ■ ■ ■ ■ ■ ■ ■ ■ ■ ■

    .005, .005

---

13. The probability that a value lies within the 95% confidence interval is
    $P = .95$. The probability that a value lies within the 99% confidence interval
    is $P = .99$. There is a greater probability of a value lying within the _____
    (95%/99%) confidence interval.

    ■ ■ ■ ■ ■ ■ ■ ■ ■ ■ ■ ■ ■ ■ ■

    99%

14. The two confidence intervals most commonly used by researchers are the
____ % and the ____ % confidence intervals.

■ ■ ■ ■ ■ ■ ■ ■ ■ ■ ■ ■ ■ ■

95, 99

---

15. Up to this point you have been considering situations where $z = (\bar{X} - \mu)/s_{\bar{X}}$ in which $s_{\bar{X}}$ represents the estimate of the population standard error of $\bar{X}$ *calculated from a large sample.* The z value thus derived has been evaluated, using Table 1, in terms of the probability values associated with the_____curve.

■ ■ ■ ■ ■ ■ ■ ■ ■ ■ ■ ■ ■

normal

---

16. For our purposes, a sample that contains approximately thirty or more measurements is called a large sample. A sample containing less than thirty measurements is called a _____ sample.

■ ■ ■ ■ ■ ■ ■ ■ ■ ■ ■ ■ ■ ■

small

---

17. Because thus far you have used Table 1 for determining probabilities, the samples from which the $s_{\bar{X}}$ has been derived have been assumed to be _____ (small/large) samples.

■ ■ ■ ■ ■ ■ ■ ■ ■ ■ ■ ■ ■ ■

large

---

18. Where the $s_{\bar{X}}$ is derived from small samples (approximately 30 or less) the probability functions of the normal curve are inaccurate for evaluating the differences between μ and $\bar{X}$. Thus, when you have only small sample data, the probabilities associated with the ratio $(\bar{X} - \mu)/s_{\bar{X}}$ _____ (can/cannot) be accurately obtained from Table 1.

■ ■ ■ ■ ■ ■ ■ ■ ■ ■ ■ ■ ■

cannot

---

19. The probabilities associated with differences between μ and $\bar{X}$ using a $s_{\bar{X}}$ estimated from a small sample, are derived from a different distribution, called the *t* distribution. The shape of the *t* distribution varies according to the number of degrees of freedom for a sample. The *df* associated with the $\bar{X}$ of a sample is $N -$ ____ .

■ ■ ■ ■ ■ ■ ■ ■ ■ ■ ■ ■ ■ ■ ■

1

20. The *t* distribution differs for each sample size, depending upon the degrees of freedom associated with the sample. Thus, the *t* distribution is _____ (the same/different) for a sample with *df* = 10 and a sample with *df* – 26.

■ ■ ■ ■ ■ ■ ■ ■ ■ ■ ■ ■ ■ ■ ■

different

Plate 68.

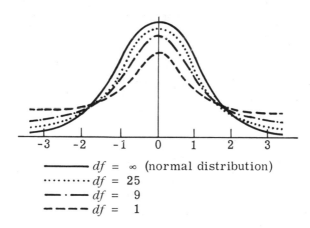

$\underline{}\ df\ =\ \infty\ \text{(normal distribution)}$
$\cdots\cdots\ df\ =\ 25$
$-\cdot-\ df\ =\ 9$
$----\ df\ =\ 1$

21. This plate presents the *t* distribution for three differing *df*'s as compared with the normal distribution. The shape of the *t* distribution which differs the most from the normal distribution is for *df* = ____ .

■ ■ ■ ■ ■ ■ ■ ■ ■ ■ ■ ■ ■ ■

one

22. **PLATE 68.** The t distribution becomes more like the normal distribution as the df of the sample _____ (increases/decreases).

■ ■ ■ ■ ■ ■ ■ ■ ■ ■ ■ ■ ■ ■

increases

23. **PLATE 68.** In this plate, the t distribution which most nearly approximates the normal distribution is the distribution for df = ____ .

■ ■ ■ ■ ■ ■ ■ ■ ■ ■ ■ ■ ■ ■

25

24. **PLATE 68.** As the df becomes very large, the t distribution becomes similar to the _____ distribution.

■ ■ ■ ■ ■ ■ ■ ■ ■ ■ ■ ■ ■ ■

normal

25. **TABLE 2.** This table presents the t values associated with each df for $P = .10$, $P = .05$, $P = .02$, and $P = .01$. From this table, the t value, when df is infinitely large (∞), is 1.960 for $P = .05$ and 2.576 for $P = .01$, which, you will recall, is the same for the _____ distribution.

■ ■ ■ ■ ■ ■ ■ ■ ■ ■ ■ ■ ■ ■

normal

26. **TABLE 2.** For $P = .05$, the t value associated with $df = 19$ is 2.093, whereas the t value associated with $df = 5$ is _____ .

■ ■ ■ ■ ■ ■ ■ ■ ■ ■ ■ ■ ■ ■

2.571

27. **TABLE 2.** As the df decreases, the t values associated with the probability levels _____ (increase/decrease).

■ ■ ■ ■ ■ ■ ■ ■ ■ ■ ■ ■ ■ ■

increase

28. TABLE 2. For a sample with $N = 18$, $df = $ _____ .

■ ■ ■ ■ ■ ■ ■ ■ ■ ■ ■ ■ ■ ■

17

---

29. TABLE 2. For a sample with $N = 18$, $df = 17$, the t value associated with $P = .05$ is _____ . The t value associated with $P = .01$ is _____ .

■ ■ ■ ■ ■ ■ ■ ■ ■ ■ ■ ■ ■ ■

2.110, 2.898

---

30. TABLE 2. To determine the 95% confidence interval for samples in which $N = 18$, you use the t values associated with $df = $ ____ , which is _____ .

■ ■ ■ ■ ■ ■ ■ ■ ■ ■ ■ ■ ■ ■

17, 2.110

---

31. Sample: $N = 18$, $df = 17$.

TABLE 2. For $P = .05$, $t = 2.110$. Whereas, for large samples, you used 1.96 $s_{\bar{X}}$ in determining the 95% confidence interval, for small samples with $N = 18$, you use _____ to determine the 95% confidence interval.

■ ■ ■ ■ ■ ■ ■ ■ ■ ■ ■ ■ ■ ■

2.110 $s_{\bar{X}}$

---

32. Sample: $N = 18$, $s_{\bar{X}} = 2$. For $P = .05$, $t = 2.110$.
Because this sample is small, the 95% confidence interval is not $\pm 1.96\, s_{\bar{X}}$, but is $\pm$ _____ $s_{\bar{X}}$.

■ ■ ■ ■ ■ ■ ■ ■ ■ ■ ■ ■ ■ ■

2.110

33. Sample: $N = 18$, $s_{\bar{X}} = 2$.
    Suppose this sample has $\bar{X} = 10$. The 95% confidence interval is designated by $\pm 2.110\, s_{\bar{X}}$. Therefore, at $-2.11\, s_{\bar{X}}$ is score value _____ , and at $2.11\, s_{\bar{X}}$ is score value _____ .

■ ■ ■ ■ ■ ■ ■ ■ ■ ■ ■ ■ ■ ■

5.78, 14.22

---

34. Sample: $N = 18$, $s_{\bar{X}} = 2$, $\bar{X} = 10$.
    At $-2.11\, s_{\bar{X}}$ is score value 5.78. At $2.11\, s_{\bar{X}}$ is score value 14.22. Therefore, the 95% confidence interval for these data is from _____ to _____ .

■ ■ ■ ■ ■ ■ ■ ■ ■ ■ ■ ■ ■ ■

5.78, 14.22

---

35. Sample: $N = 18$, $s_{\bar{X}} = 2$, $\bar{X} = 10$. The 95% confidence interval is from 5.78 to 14.22. This means that you could say the probability is _____ that intervals derived in this manner from a series of samples of this size would contain μ.

■ ■ ■ ■ ■ ■ ■ ■ ■ ■ ■ ■ ■ ■

.95

---

36. The 99% confidence interval derived from small sample statistics may be determined in exactly the same manner as the 95% confidence interval by using Table 2 to determine the t value in the column headed $P =$ _____ opposite the df of the sample.

■ ■ ■ ■ ■ ■ ■ ■ ■ ■ ■ ■ ■ ■

.01

---

37. Inspection of Table 2 indicates that for an infinite size sample (where $df = \infty$), at $P = .05$, $t =$ ___. This is identical to the z-score determined earlier for the _____ distribution.

■ ■ ■ ■ ■ ■ ■ ■ ■ ■ ■ ■ ■ ■

1.960, normal

38. Thus, in theory, for infinitely large samples, the t distribution ___ (is/is not) identical to the normal distribution.

■ ■ ■ ■ ■ ■ ■ ■ ■ ■ ■ ■ ■

is

39. In practice, for samples with degrees of freedom greater than thirty, the normal distribution is a good approximation of the ___ (symbol) distribution.

■ ■ ■ ■ ■ ■ ■ ■ ■ ■ ■ ■ ■

t

40. Because the size of t increases as the df of the sample decreases, it follows that the confidence interval is _____ (larger/smaller) for small samples than for large samples.

■ ■ ■ ■ ■ ■ ■ ■ ■ ■ ■ ■ ■

larger

EXERCISES

1. What are the two confidence intervals generally employed by researchers? Which is the more stringent?

2. Determine the 99% confidence interval for the following samples:
   (a) $N = 121, s = 7.4, \bar{X} = 100$.
   (b) $N = 64, s = 5.2, \bar{X} = 75$.

3. Determine the 95% and 99% confidence intervals for the following samples:
   (a) $N = 20, s_{\bar{X}} = 3.5, \bar{X} = 50$.
   (b) $N = 12, s_{\bar{X}} = 4.11, \bar{X} = 32.6$.

4. What is the t distribution? How is it related to sample size?

NULL HYPOTHESIS /
THE STANDARD ERROR OF THE
DIFFERENCE BETWEEN MEANS

Previous sets have explored statistical methods for determining various confidence intervals. Another important function of statistics is in estimating the validity of specific predictions or hypotheses. Such hypothesis–testing involves statistical processes. For example, a researcher may wish to discover whether a certain treatment has a significant effect upon the behavior of people; or whether, under specified conditions, certain changes take place. In any case, he generally makes a prediction. To evaluate whether he should accept or reject this prediction, he states his prediction as a hypothesis, which he then attempts to prove or disprove by statistical decision-making processes.

The hypothesis that is used when making a statistical analysis of data is somewhat different from the researcher's hypothesis. This set will present the definition of a *null hypothesis* and explain how to determine whether to accept or reject it. It will also discuss sampling error in connection with the means of two samples. A new term, the *standard error of the difference between means,* will be defined and a method will be given for determining whether to accept or reject the null hypothesis on the basis of data from two samples.

SPECIFIC OBJECTIVES OF SET 14

At the conclusion of this set you will be able to:

(1) state a research hypothesis as a null hypothesis.

(2) state what is meant by the *standard error of the difference between means.*

(3) make a determination about whether to accept or reject the null hypothesis, based upon the probability that the difference between two sample means is due to sampling error.

(4) identify the symbol $\sigma_{\bar{X}_1 - \bar{X}_2}$.

1. Statistical decision making is generally concerned with determining probability levels associated with hypotheses. The statement "The sample $\overline{X}$ lies between score value 30 and score value 50" could be taken as a _____.

■ ■ ■ ■ ■ ■ ■ ■ ■ ■ ■ ■ ■ ■

hypothesis

2. If you have data for one sample drawn from a population, you can make probability statements regarding the acceptability of a _____ about the population.

■ ■ ■ ■ ■ ■ ■ ■ ■ ■ ■ ■ ■ ■

hypothesis

3. You decide whether or not to accept a hypothesis by determining its _____ of being true.

■ ■ ■ ■ ■ ■ ■ ■ ■ ■ ■ ■ ■ ■

probability

4. When you are concerned with whether or not two populations differ, you must make a _____ regarding the difference between the two populations.

■ ■ ■ ■ ■ ■ ■ ■ ■ ■ ■ ■ ■

hypothesis

5. Statistical hypotheses are generally stated in *null* form. The word *null* means *no*. Thus, a null hypothesis states there is _____ difference between two populations.

■ ■ ■ ■ ■ ■ ■ ■ ■ ■ ■ ■ ■ ■

no

6   Here is a null hypothesis: "There is no difference between the heights of tenth-grade boys and tenth-grade girls." This is the same as saying "The difference between the heights of tenth-grade boys and girls is equal to _____."

■ ■ ■ ■ ■ ■ ■ ■ ■ ■ ■ ■ ■

zero

---

7. A hypothesis that states there is no difference between two populations is called a _____ hypothesis.

■ ■ ■ ■ ■ ■ ■ ■ ■ ■ ■ ■ ■

null

---

8. If there is no difference between the heights of tenth-grade boys and girls, you may say that, in regard to height, they come from the same _____ .

■ ■ ■ ■ ■ ■ ■ ■ ■ ■ ■ ■ ■

population

---

9. Because it is not practical to measure the height of all tenth-grade boys and girls, you would probably measure a _____ of each group.

■ ■ ■ ■ ■ ■ ■ ■ ■ ■ ■ ■ ■

sample

---

10. If there is no difference between the $\bar{X}$'s of the two samples, you can conclude that there is _____ difference between the heights of the populations of tenth-grade boys and tenth-grade girls.

■ ■ ■ ■ ■ ■ ■ ■ ■ ■ ■ ■ ■

no

---

11. You recall, however, that the $\bar{X}$'s of two samples will seldom be *exactly* the same, due to _____ error.

■ ■ ■ ■ ■ ■ ■ ■ ■ ■ ■ ■ ■

sampling

---

12. The $\bar{X}$'s of two samples may differ somewhat, due to sampling error, even when the samples are selected from the same _____ .

■ ■ ■ ■ ■ ■ ■ ■ ■ ■ ■ ■ ■ ■

population

---

13. TABLE 2. In addition to the t values for $P = .05$, the t values for $P = $_____ are also presented in the fourth column of this table.

■ ■ ■ ■ ■ ■ ■ ■ ■ ■ ■ ■ ■ ■

.01

---

14. TABLE 2. The second and fourth column of this table present the t values for $P = $_____ and $P = $_____ for the differing df, because these two probability levels are those generally used in reporting research findings.

■ ■ ■ ■ ■ ■ ■ ■ ■ ■ ■ ■ ■ ■

.05, .01

---

15. When you *test* the null hypothesis, you are asking this question: "Is the difference between the $\bar{X}$'s of the two samples larger than can be expected due to sampling _____ ?"

■ ■ ■ ■ ■ ■ ■ ■ ■ ■ ■ ■ ■ ■

error

---

16. If the difference between the $\bar{X}$'s of two samples is larger than can be expected as a result of sampling error, then you reject the null hypothesis and conclude that the two samples were selected from two different _____ .

■ ■ ■ ■ ■ ■ ■ ■ ■ ■ ■ ■ ■ ■

populations

---

17. If you selected a pair of samples from a population, it is likely that, due to sampling error, the $\bar{X}$'s of the pair of samples would _____ .

■ ■ ■ ■ ■ ■ ■ ■ ■ ■ ■ ■ ■ ■

differ

185

18. If you selected a large number of pairs of samples, you could obtain a $\overline{X}$ difference score for each _____ of samples.

■ ■ ■ ■ ■ ■ ■ ■ ■ ■ ■ ■ ■ ■

pair

---

19. The $\overline{X}$ difference score for two samples would be obtained by subtracting the $\overline{X}$ of the first sample from the ___ (symbol) of the _____ sample.

■ ■ ■ ■ ■ ■ ■ ■ ■ ■ ■ ■ ■ ■

$\overline{X}$, second

---

20. Thus, if you subtracted the $\overline{X}$ of the second sample from the $\overline{X}$ of the first sample, the $\overline{X}$ difference score would be positive if the _____ (first/second) sample $\overline{X}$ was larger than the _____ (first/second) sample $\overline{X}$.

■ ■ ■ ■ ■ ■ ■ ■ ■ ■ ■ ■ ■ ■

first, second

---

21. Therefore, in many pairs of samples, if you had randomly paired your samples, you should have just as many positive $\overline{X}$ difference scores as you would have _____ $\overline{X}$ difference scores.

■ ■ ■ ■ ■ ■ ■ ■ ■ ■ ■ ■ ■ ■

negative

---

22. The standard deviation of sample means, you recall, is called the *standard error of the mean*. Therefore, the standard deviation of the *difference* between sample $\overline{X}$'s is called the *standard error of the* _____ .

■ ■ ■ ■ ■ ■ ■ ■ ■ ■ ■ ■ ■

difference

---

23. The standard error of the difference is the standard deviation of a distribution of differences between sample ___'s (symbol).

■ ■ ■ ■ ■ ■ ■ ■ ■ ■ ■ ■ ■

$\overline{X}$

Plate 69. Sampling Distribution of Differences Between Pairs of Sample $\overline{X}$'s for All Possible Pairs of Samples

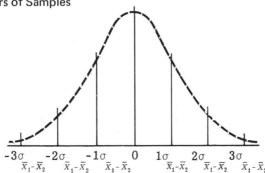

24. As indicated in this plate, the symbol for the standard error of the difference is $\sigma_{\overline{X}_1 - \overline{X}_2}$, which represents a standard deviation of $\overline{X}$ _____ scores.

■ ■ ■ ■ ■ ■ ■ ■ ■ ■ ■ ■ ■

difference

25. PLATE 69. The $\overline{X}$ of this distribution is zero. This indicates that the $\overline{X}$ of all the difference scores derived from pairs of samples selected from one population is _____.

■ ■ ■ ■ ■ ■ ■ ■ ■ ■ ■ ■ ■

zero

26. PLATE 69. The symbol $\sigma_{\overline{X}_1 - \overline{X}_2}$ indicates the standard error of the difference, which is the _____ _____ of differences between the $\overline{X}$'s of all possible pairs of same-sized samples.

■ ■ ■ ■ ■ ■ ■ ■ ■ ■ ■ ■ ■

standard deviation

27. PLATE 69. The symbol for the standard error of the difference is _____ .

■ ■ ■ ■ ■ ■ ■ ■ ■ ■ ■ ■ ■

$\sigma_{\overline{X}_1 - \overline{X}_2}$

187

28. PLATE 69. The probability functions of the normal curve apply to this particular distribution of mean differences, because the $\sigma_{\overline{X}_1 - \overline{X}_2}$ is assumed to be computed from the differences between the $\overline{X}$'s of all possible pairs (i.e., the population) of samples. Thus, it is here considered that the standard error of the difference is a _____ (statistic/parameter) and it is assigned a _____ (Roman/Greek) symbol.

■ ■ ■ ■ ■ ■ ■ ■ ■ ■ ■ ■ ■ ■

parameter, Greek

Note: It is usually impossible to obtain all the possible pairs of samples and derive $\sigma_{\overline{X}_1 - \overline{X}_2}$ from them. However, it is possible to obtain two samples and from them make an *estimate* of the standard error of the difference. The symbol for the estimate of the standard error of the difference is $s_{\overline{X}_1 - \overline{X}_2}$. The method for making this estimate is presented in Set 15.

---

29. The probability values associated with differences between two sample means, when $s_{\overline{X}_1 - \overline{X}_2}$ has been computed from the two samples, does not follow the normal distribution when the samples are small. As was the case when considering $s_{\overline{X}}$, with small samples, the ratio between the mean difference and $s_{\overline{X}_1 - \overline{X}_2}$ is evaluated according to the probability values associated with its particular ____ distribution.

■ ■ ■ ■ ■ ■ ■ ■ ■ ■ ■ ■ ■ ■

t

---

30. The shape of the *t* distribution curve varies with the number of _____ of _____ available.

■ ■ ■ ■ ■ ■ ■ ■ ■ ■ ■ ■ ■ ■

degrees, freedom

---

31. PLATE 69. If the shape of the curve of the *t* distribution varies with the number of *df* available, then the proportions of the area under the curve also _____ with the number of *df* available.

■ ■ ■ ■ ■ ■ ■ ■ ■ ■ ■ ■ ■ ■

vary

---

32. If you reject the null hypothesis, you conclude that the two samples were selected from _____ (the same/different) population(s).

■ ■ ■ ■ ■ ■ ■ ■ ■ ■ ■ ■ ■ ■

different

33. If you accept the null hypothesis, you conclude, from the obtained sample statistics, that there is insufficient evidence to _____ it.

■ ■ ■ ■ ■ ■ ■ ■ ■ ■ ■ ■ ■ ■

reject

34. In making a decision whether to accept or reject the null hypothesis, you must first assume that the two sample $\overline{X}$'s come from the same population. When you make this assumption you are saying that the difference between the $\overline{X}$'s is only due to _____ error.

■ ■ ■ ■ ■ ■ ■ ■ ■ ■ ■ ■ ■ ■

sampling

35. Suppose you decide to use $P = .05$ as the "cut-off" point in accepting or rejecting the _____ hypothesis.

■ ■ ■ ■ ■ ■ ■ ■ ■ ■ ■ ■ ■ ■

null

36. Assume that you have $\overline{X}$'s of two samples. If the probability of obtaining a difference as large as you obtain between your samples is $P = .05$ or less, there is a very small probability that the two samples are from the same population. Therefore, you should _____ (accept/reject) the null hypothesis.

■ ■ ■ ■ ■ ■ ■ ■ ■ ■ ■ ■ ■ ■

reject

37. If the probability of obtaining a difference as large as you obtain between your samples is greater than $P = .05$, there is a large probability that the two samples are from the same population and that the difference is due only to sampling error. Therefore, you should _____ (accept/reject) the null hypothesis.

■ ■ ■ ■ ■ ■ ■ ■ ■ ■ ■ ■ ■

accept

Plate 70. t Distribution of $\bar{X}$ Difference Scores

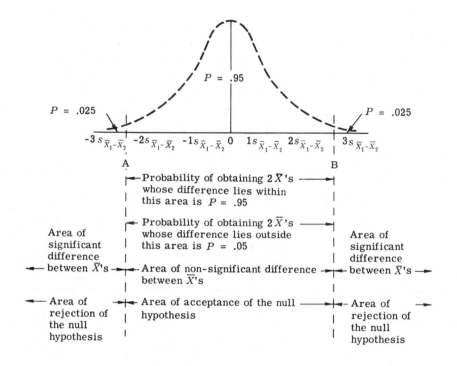

38. This plate presents a diagram depicting the t distribution. The curve of the distribution is indicated by a dotted line because the actual shape of the t distribution depends upon the ___ (symbol).

■ ■ ■ ■ ■ ■ ■ ■ ■ ■ ■ ■ ■

df

190

39. PLATE 70. Note points A and B in this plate. The probability of obtaining a $\overline{X}$ difference score that lies within the area designated by these two points is $P =$ _____ .

.95

---

40. PLATE 70. The probability of obtaining a $\overline{X}$ difference score that lies below the point designated by A is $P =$ _____ .

.025

---

41. PLATE 70. The probability of obtaining a $\overline{X}$ difference score that lies above the point designated by B is $P =$ _____ .

.025

---

42. PLATE 70. The probability of obtaining a $\overline{X}$ difference score that lies either above or below the area between points A and B is $P =$ _____ .

.05   (.025 + .025)

---

43. PLATE 70. The probability is so very small $(P = .05)$ of your obtaining a $\overline{X}$ difference score that lies outside the area between points A and B that researchers are willing to _____ (accept/reject) the null hypothesis if they obtain a difference that lies outside this area.

reject

---

44. If you accept the null hypothesis, you are in effect saying that the $\overline{X}$ difference you obtained is due to sampling error and the difference is therefore _____ (significant/non-significant).

non-significant

---

45. If you reject the null hypothesis, you are in effect saying that the $\overline{X}$ difference score that you obtained is so large that it is not due *only* to sampling error, and therefore the difference is _____ (significant/non-significant).

■ ■ ■ ■ ■ ■ ■ ■ ■ ■ ■ ■ ■ ■

significant

46. PLATE 70. If you reject the null hypothesis, you are saying that the difference between your sample $\overline{X}$'s is significant. This is the same as concluding that the two samples come from different _____ .

■ ■ ■ ■ ■ ■ ■ ■ ■ ■ ■ ■ ■ ■

populations

47. PLATE 70. The lines drawn at points A and B are dotted, which indicates that their placement along the base line varies with the shape of the distribution curve, depending upon the ____ (symbol).

■ ■ ■ ■ ■ ■ ■ ■ ■ ■ ■ ■ ■ ■

df

EXERCISES

1. A large number of college freshmen are randomly assigned to two groups which are to be taught elementary statistics by different methods. A final examination will be given. State the null hypothesis to be tested.

2. A researcher makes the hypothesis that children who study in the morning should receive higher algebra test scores than children who study in the evening. A final examination is given at the end of the semester. State the null hypothesis to be tested.

3. What is meant by the term *standard error of the difference*?

4. If you designate $P = .05$ as your acceptable level for significance, and your research study yields a significance level of $P = .13$, would you accept or reject the null hypothesis?

5. When you accept the null hypothesis, what statement can you make about the difference between your two sample means?

6. What does the symbol $\sigma_{\bar{X}_1 - \bar{X}_2}$ represent?

t RATIO FOR INDEPENDENT MEANS

In the previous set a method was presented for accepting or rejecting the null hypothesis on the basis of the standard error of the difference between pairs of sample means. However, it is rarely possible to select a large number of samples, determine the difference between their means, and distribute these differences. A method has therefore been devised for estimating the standard error of the difference between means from the data in just two samples.

This set will present a method for "pooling" sums of squares in order to estimate the standard error of the difference between means, and also for determining whether or not two samples differ significantly by employing the t test of the significance between sample means.

The statistical concept of "significant difference" will also be defined.

SPECIFIC OBJECTIVES OF SET 15

At the conclusion of this set you will be able to:

(1) using Formula 15, estimate the standard error of the difference from data contained in two samples.

(2) conduct a t test for the significance of the difference between two sample means, using Formula 16.

(3) using Table 2, evaluate the level of significance of a t ratio.

(4) determine whether to accept or reject the null hypothesis on the basis of the t test.

(5) identify the symbol $s_{\overline{X}_1 - \overline{X}_2}$.

1. If you had many samples selected from a population, you could compute the difference between the $\overline{X}$'s for each pair of samples, and then calculate the standard deviation of the $\overline{X}$ difference scores. This would be called the standard error of the _____, and its symbol would be _____ .

■ ■ ■ ■ ■ ■ ■ ■ ■ ■ ■ ■ ■ ■

difference, $\sigma_{\overline{X}_1 - \overline{X}_2}$

---

2. When you have only two samples, you may use the data contained in them to estimate the standard error of the difference. Since the standard error term is derived from sample data, its formula is logically written _____ $\overline{X}_1 - \overline{X}_2$.

■ ■ ■ ■ ■ ■ ■ ■ ■ ■ ■ ■ ■ ■

s

To compute $s_{\overline{X}_1 - \overline{X}_2}$ you first need to obtain s^2.

Formula 14. Estimation of the Common Population Variance (Pooled Variance) from the Data in Two Samples (Formulas 14a and 14b are Equivalent)

$$s^2 = \frac{\Sigma x_1^2 + \Sigma x_2^2}{N_1 + N_2 - 2} \qquad \text{(Formula 14a)}$$

$$s^2 = \frac{(N_1 - 1)s_1^2 + (N_2 - 1)s_2^2}{N_1 + N_2 - 2} \qquad \text{(Formula 14b)}$$

---

3. FORMULA 14a. In this formula, Σx_1^2 indicates the sum of squares for the first sample, and Σx_2^2 indicates the sum of squares for the _____ sample.

■ ■ ■ ■ ■ ■ ■ ■ ■ ■ ■ ■ ■ ■

second

---

4. FORMULA 14a. The symbol N_1 indicates the number of scores in the _____ (first/second) sample. The symbol N_2 indicates the number of scores in the _____ (first/second) sample.

■ ■ ■ ■ ■ ■ ■ ■ ■ ■ ■ ■ ■

first, second

---

5. FORMULA 14a. The part of the formula that states $\Sigma x_1^2 + \Sigma x_2^2$ indicates that you "pool" the sum of squares of the two _____ .

■ ■ ■ ■ ■ ■ ■ ■ ■ ■ ■ ■ ■

samples

---

6. FORMULA 14a. The part of the formula that states $N_1 + N_2 - 2$ indicates that you "pool" the degrees of _____ of the two samples.

Hint: $(N_1 - 1) + (N_2 - 1) = N_1 + N_2 - 2$

■ ■ ■ ■ ■ ■ ■ ■ ■ ■ ■ ■ ■

freedom

---

7. Recall from Formula 12 that when you divide the sum of squares by df, you obtain an estimate of the population _____ .

■ ■ ■ ■ ■ ■ ■ ■ ■ ■ ■ ■ ■

variance

---

8. When you divide Σx^2 by df, you obtain the variance estimate. Thus, the part of Formula 14a that states

$$\frac{\Sigma x_1^2 + \Sigma x_2^2}{N_1 + N_2 - 2}$$

indicates that you are "pooling" the _____ estimates of the two samples.

■ ■ ■ ■ ■ ■ ■ ■ ■ ■ ■ ■ ■

variance

---

9. FORMULA 14b. This formula is equivalent to Formula 14a and indicates that you are estimating the population variance by _____ the variance estimates of two samples.

■ ■ ■ ■ ■ ■ ■ ■ ■ ■ ■ ■ ■ ■

pooling

Plate 71. Height of Tenth-Grade Children

1st Sample Boys	2nd Sample Girls
$N_1 = 26$	$N_2 = 37$
$\overline{X}_1 = 66$ inches	$\overline{X}_2 = 67$ inches
$\Sigma X_1 = 1716$	$\Sigma X_2 = 2479$
$\Sigma X_1^2 = 113{,}338$	$\Sigma X_2^2 = 166{,}205$

10. This plate gives the data for two samples: the heights of tenth-grade boys and the heights of tenth-grade girls. The null hypothesis that is to be tested is: There is _____ difference between the heights of tenth-grade boys and tenth-grade girls.

■ ■ ■ ■ ■ ■ ■ ■ ■ ■ ■ ■ ■ ■

no

---

11. PLATE 71. The null hypothesis to be tested is: There is no difference between the heights of tenth-grade boys and tenth-grade girls. This is the same as saying that the true difference between the $\overline{X}$ of these two groups is

_____ .

■ ■ ■ ■ ■ ■ ■ ■ ■ ■ ■ ■ ■ ■

zero

---

12. PLATE 71. The data indicate that the difference between the means of these two samples is ___ inch(es).

■ ■ ■ ■ ■ ■ ■ ■ ■ ■ ■ ■ ■ ■

1

---

13. PLATE 71. The question to be answered is: How probable is it that the difference of 1 inch between the sample $\overline{X}$s is due to sampling _____ ?

■ ■ ■ ■ ■ ■ ■ ■ ■ ■ ■ ■ ■ ■

error

---

14. When you determine the probability that your sample $\overline{X}$ difference is due to sampling error, you are testing the _____ hypothesis.

■ ■ ■ ■ ■ ■ ■ ■ ■ ■ ■ ■ ■ ■

null

---

15. To test the null hypothesis, Formula 14 indicates that you must first determine an estimate of the _____ population variance, derived from the data in the two samples.

■ ■ ■ ■ ■ ■ ■ ■ ■ ■ ■ ■ ■ ■

common

---

16. PLATE 71. Using Formula 11b for the computation of Σx^2 by the raw score method, for each sample substitute the appropriate values for the symbols and do the necessary arithmetic:

Boys Σx_1^2 = _____

Girls Σx_2^2 = _____

■ ■ ■ ■ ■ ■ ■ ■ ■ ■ ■ ■ ■ ■

$$\text{Boys } \Sigma x_1^2 = 113338 - \frac{(1716)^2}{26} = 82$$

$$\text{Girls } \Sigma x_2^2 = 166205 - \frac{(2479)^2}{37} = 112$$

17. Formula 15 indicates that you obtain an estimate of the standard error of the difference between means by using the _____ variance estimate, as derived in Formula 14.

■ ■ ■ ■ ■ ■ ■ ■ ■ ■ ■ ■ ■ ■

pooled

18. PLATE 71. Boys $\Sigma x_1^2 = 82$, Girls $\Sigma x_2^2 = 112$. To estimate the common population variance, substitute the appropriate values for the symbols in Formula 14a.

$$s^2 = \frac{\rule{3cm}{0.4pt}}{\rule{3cm}{0.4pt}}$$

■ ■ ■ ■ ■ ■ ■ ■ ■ ■ ■ ■ ■ ■

$$\frac{82 + 112}{26 + 37 - 2}$$

19. PLATE 71. In this example, the pooled variance estimate, as derived using Formula 14a, is $s^2 = 3.18$. Using Formula 15, substitute the values in the following:

$$s_{\bar{X}_1 - \bar{X}_2} = \sqrt{\frac{\rule{2cm}{0.4pt}}{\rule{2cm}{0.4pt}} + \frac{\rule{2cm}{0.4pt}}{\rule{2cm}{0.4pt}}}$$

■ ■ ■ ■ ■ ■ ■ ■ ■ ■ ■ ■ ■ ■

$$\sqrt{\frac{3.18}{26} + \frac{3.18}{37}}$$

20. PLATE 71. Do the necessary arithmetic to determine the value of $s_{\overline{X}_1 - \overline{X}_2}$.

$$s_{\overline{X}_1 - \overline{X}_2} = \underline{\hspace{2cm}} \text{(to three decimal places)}$$

■ ■ ■ ■ ■ ■ ■ ■ ■ ■ ■ ■ ■

.455

---

21. PLATE 71. The difference between the $\overline{X}$ heights of boys and girls is __ .

■ ■ ■ ■ ■ ■ ■ ■ ■ ■ ■ ■ ■

1

---

22. To determine the probability of obtaining a difference score of 1, you must compute a *t* ratio. Formula 16 gives the formula for determining the *t* ratio.

Formula 16. Calculation of the *t* Ratio for Independent Means

$$t = \frac{\overline{X}_1 - \overline{X}_2}{s_{\overline{X}_1 - \overline{X}_2}}$$

Degrees of Freedom: $N_1 + N_2 - 2$

The symbol *t* is used in this formula because the two samples have _____ (small/large) *N*'s.

■ ■ ■ ■ ■ ■ ■ ■ ■ ■ ■ ■ ■

small

---

23. FORMULA 16. The title of this formula indicates that it is for _____ samples. The meaning of this term will be fully discussed later.

■ ■ ■ ■ ■ ■ ■ ■ ■ ■ ■ ■ ■

independent

---

24. FORMULA 16. This is called a t ratio because it is a ratio between the difference of the two sample $\bar{X}$s and the _____ _____ of the
_____ .

■ ■ ■ ■ ■ ■ ■ ■ ■ ■ ■ ■ ■ ■

standard error, difference

25. PLATE 71. You determined that $s_{\bar{X}_1-\bar{X}_2} = .455$. Using Formula 16, calculate the t ratio by substituting the numerical values for the symbols and doing the necessary arithmetic.

■ ■ ■ ■ ■ ■ ■ ■ ■ ■ ■ ■ ■ ■

$$t = \frac{67 - 66}{.455} = \frac{1}{.455} = 2.197$$

26. PLATE 71. $t = 2.197$
To interpret the meaning of this t ratio, you must first determine the df associated with it. The df for the boy sample is $N_1 - 1$, or _____ . The df for the girl sample is $N_2 - 1$, or _____ .

■ ■ ■ ■ ■ ■ ■ ■ ■ ■ ■ ■ ■ ■

25, 36

27. As shown below Formula 16, the degrees of freedom associated with this t-ratio is $N_1 + N_2 - 2$. This is algebraically the same as $(N_1 - 1) + (N_2 - 1)$ which is simply the sum of the _____ 's (symbol) associated with the two samples.

■ ■ ■ ■ ■ ■ ■ ■ ■ ■ ■ ■ ■ ■

df

28. PLATE 71. $t = 2.197, N_1 = 26, N_2 = 37$.
$df = N_1 + N_2 - 2$. For these data, $df =$ _____ .

■ ■ ■ ■ ■ ■ ■ ■ ■ ■ ■ ■ ■ ■

61 $(26 + 37 - 2)$

29. The $\bar{X}$s of two samples are said to be significantly different (and hence, come from different populations) if the probability of obtaining the difference that you did is equal to or less than $P = .05$. Table 2 presents the t ratios corresponding to $P =$ _____ , _____ , _____ , and _____ , for differing df's.

■ ■ ■ ■ ■ ■ ■ ■ ■ ■ ■ ■ ■ ■

.10, .05, .02, .01

---

30. TABLE 2. The values of t required to reach the probability level of $P = .05$ are presented because $P = .05$ is the largest probability that most statisticians will accept as indicating a significant difference between sample ___'s (symbol).

■ ■ ■ ■ ■ ■ ■ ■ ■ ■ ■ ■ ■ ■

$\bar{X}$

---

31. TABLE 2. If an obtained t ratio is larger than the tabled value of t for $P = .05$, you conclude that the two samples have been selected from different _____ .

■ ■ ■ ■ ■ ■ ■ ■ ■ ■ ■ ■ ■ ■

populations

---

32. In testing the null hypothesis, if the t ratio is larger than the tabled value of t at $P = .05$, the difference between the two samples is said to be _____ (significant/non-significant).

■ ■ ■ ■ ■ ■ ■ ■ ■ ■ ■ ■ ■ ■

significant

---

33. If your t ratio indicates that the $\bar{X}$s of the samples are significantly different, you reject the null hypothesis, because the null hypothesis states that the two samples come from the same _____ .

■ ■ ■ ■ ■ ■ ■ ■ ■ ■ ■ ■ ■ ■

population

---

34. PLATE 71. $t = 2.197, df = 61$.
From Table 2, the value of t required for $P = .05$ with $df = 61$ is _____ .
(Use the nearest listed df in the table.)

■ ■ ■ ■ ■ ■ ■ ■ ■ ■ ■ ■ ■ ■ ■

2.000
(the nearest listed df is for $df = 60$)

Plate 72. t Distribution for $df = 61$

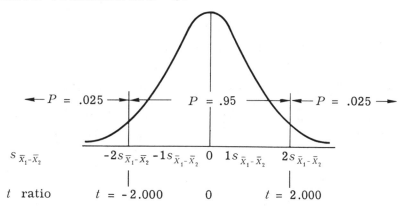

35. This plate presents the t distribution for $df = 61$, which is the distribution appropriate for Plate 71. Notice that the probability at either extreme of the distribution is $P =$ _____ . The sum of these two probabilities is $P =$ ____ .

■ ■ ■ ■ ■ ■ ■ ■ ■ ■ ■ ■ ■

.025, .05

---

36. PLATE 72. For $df = 61$, the probability of obtaining a t ratio *above* 2.000 is $P =$ _____ .

■ ■ ■ ■ ■ ■ ■ ■ ■ ■ ■ ■ ■

.025

---

37. PLATE 72. For $df = 61$, the probability of obtaining a t ratio *below* $t = -2.000$ is $P =$ _____ .

■ ■ ■ ■ ■ ■ ■ ■ ■ ■ ■ ■ ■

.025

203

38. PLATE 72. For $df = 61$, the probability of obtaining a t ratio either above $t = 2.000$ or below $t = -2.000$ is $P =$ _____ .

■ ■ ■ ■ ■ ■ ■ ■ ■ ■ ■ ■ ■ ■

.05

---

39. PLATE 72. For $df = 61$, the probability of obtaining a t ratio below $t = -2.000$ or above $t = 2.000$ is $P = .05$. Therefore, for this df the probability of a $\overline{X}$ difference score yielding a t ratio as large as either plus or minus 2.000 is _____ .

■ ■ ■ ■ ■ ■ ■ ■ ■ ■ ■ ■ ■ ■

.05

---

40. PLATE 71. $t = 2.197, df = 61$.
From Table 2, the t ratio required for $P = .05$ is 2.000. The t ratio obtained from the data in Plate 71 ___ (is/is not) as large as or larger than is required for $P = .05$.

■ ■ ■ ■ ■ ■ ■ ■ ■ ■ ■ ■ ■ ■

is

---

41. PLATE 71. $t = 2.197, df = 61$.
This t ratio is larger than $t = 2.000$, which is required for significance at $P = .05$. Therefore, you _____ (accept/reject) the null hypothesis that there is no difference between the populations from which these two samples were selected.

■ ■ ■ ■ ■ ■ ■ ■ ■ ■ ■ ■ ■ ■

reject

---

42. If the obtained t ratio had been less than 2.000, you would conclude that, from these sample statistics, you could not _____ (accept/reject) the hypothesis that these two sample means came from the same population.

■ ■ ■ ■ ■ ■ ■ ■ ■ ■ ■ ■ ■ ■

reject

EXERCISES

1. Calculate the standard error of the difference, using Formula 15, and the t ratio, using Formula 16, for the means of the following two groups.

Group A	Group B
$N = 13$	$N = 19$
$\bar{X} = 7$	$\bar{X} = 9.4$
$\Sigma x^2 = 96$	$\Sigma x^2 = 112$

2. Using Table 2, evaluate the significance of the t ratio obtained in Exercise 1. Would you reject the null hypothesis at $P = .05$? At $P = .01$?

3. For the following two groups, calculate the standard error of the difference, using Formula 15. Using Formula 16, calculate the t ratio.

Group X	Group Y
$N = 11$	$N = 17$
$\bar{X} = 7.6$	$\bar{X} = 10.9$
$\Sigma x^2 = 77$	$\Sigma x^2 = 98$

4. Using Table 2, evaluate the significance of the t ratio obtained in Exercise 3. Would you reject the null hypothesis at $P = .05$? At $P = .01$?

5. Identify the symbol $s_{\bar{X}_1 - \bar{X}_2}$.

t RATIO FOR NON-INDEPENDENT MEANS

The *t* ratio was discussed in Set 15 as a method of determining the significance of the difference between sample means; it tested the difference between what may be termed *independent*, or uncorrelated, samples. When making mean score comparisons, however, there is sometimes a correlation between the two sets of scores being examined. This may occur when the two sets of scores are derived from the same sample. For instance, when a sixth-grade class is given a spelling test at the beginning and end of the school year, the two sets of scores are correlated because they are derived from the same sample of children. Correlated sets of scores also occur when individuals in two samples are "matched" on a third variable, such as chronological age. The means of such samples are said to be *non-independent* because of this matching process. This is so because each individual in one sample has been matched with an individual in the other sample. The concept of correlation will be examined in Set 21. This set presents a method for computing the *t* ratio for non-independent means and evaluating it for statistical significance.

SPECIFIC OBJECTIVES OF SET 16

At the conclusion of this set you will be able to:

(1) distinguish between independent and non-independent sets of scores.

(2) use Formulas 17 through 19 to calculate a *t* ratio for data of two non-independent samples and evaluate it for level of significance.

(3) identify the symbol D.

1. You have already been given the formulas for the t test for the difference between independent means (see Formula 16). The word "independent" means that the sets of scores _____ (are/are not) related.

■ ■ ■ ■ ■ ■ ■ ■ ■ ■ ■ ■ ■ ■

are not

2. When the scores on two variables are related, they are said to be _____(independent/non-independent).

■ ■ ■ ■ ■ ■ ■ ■ ■ ■ ■ ■ ■ ■

non-independent

3. If you obtain arithmetic scores on two different samples of individuals, and if the two samples are randomly selected, the two sets of scores are _____.

■ ■ ■ ■ ■ ■ ■ ■ ■ ■ ■ ■ ■ ■

independent

4. If you match the individuals in the two samples on some variable, say chronological age or mental age, then the individuals in the two samples are said to be _____.

■ ■ ■ ■ ■ ■ ■ ■ ■ ■ ■ ■ ■ ■

non-independent

5. If you obtain test scores on each student in one sample, at the beginning and end of an instructional period, the scores you obtain on the first testing are _____ (independent/non-independent) of the scores obtained on the second testing because the two sets of data are on _____ (the same/different) students.

■ ■ ■ ■ ■ ■ ■ ■ ■ ■ ■ ■ ■ ■

non-independent, the same

6. In the case above, you would need to perform a t test of the significance between _____ means.

non-independent

To illustrate the method for performing a t test between non-independent means, consider the following example:

Plate 73.

Individual	First Testing X_1	Second Testing X_2	D	D^2
A	4	2	2	4
B	4	2	2	4
C	3	1	2	4
D	3	4	−1	1
E	5	4	1	1
F	2	1	1	1
G	4	3	1	1
H	3	3	0	0
I	2	2	0	0
J	1	3	−2	4
$N = 10$	$\Sigma X_1 = 31$	$\Sigma X_2 = 25$	$\Sigma D = 6$	$\Sigma D^2 = 20$

7. In this plate, individual A received a score on the first testing of 4, and a score on the second testing of ___.

2

8. PLATE 73. Individual A has a difference score (D) from the first testing to the second testing of $X_1 - X_2 = 4 - 2 =$ ___.

2

9. PLATE 73. For individual **A**, $D = 2$ because his score value decreased from the first testing to the second testing. For individual **J**, $D =$ ____ .

■ ■ ■ ■ ■ ■ ■ ■ ■ ■ ■ ■ ■ ■

-2

10. PLATE 73. For individual J, $D = -2$ because his score value _____ (increased/decreased) from the first testing to the second testing.

■ ■ ■ ■ ■ ■ ■ ■ ■ ■ ■ ■ ■

increased

11. PLATE 73. The third column of this plate presents D for each of the pairs of scores. The fourth column presents _____(symbol) for each of the pairs of scores.

■ ■ ■ ■ ■ ■ ■ ■ ■ ■ ■ ■ ■ ■

D^2

Formula 17. Estimation of the Population Variance of Difference Scores

$$s_D^2 = \frac{N\Sigma D^2 - (\Sigma D)^2}{N(N-1)}$$

where $D = X_1 - X_2$ for each pair of scores

---

12. In Formula 17, N is the number of pairs of scores. For the example in Plate 73, $N =$ ____ .

■ ■ ■ ■ ■ ■ ■ ■ ■ ■ ■ ■ ■ ■

10

13. Formula 17. In order to determine s_D^2 for the example in Plate 73, determine $\Sigma D^2 =$ ____ , $(\Sigma D)^2 =$ ____ .

■ ■ ■ ■ ■ ■ ■ ■ ■ ■ ■ ■ ■ ■

20, 36

14. **PLATE 73.** Using Formula 17, substitute the values and solve for s_D^2.

$$s_D^2 = \underline{\hspace{1cm}}\text{(to two decimal places)}$$

■ ■ ■ ■ ■ ■ ■ ■ ■ ■ ■ ■ ■ ■

$$\frac{10(20) - (6)^2}{10(9)} = 1.82$$

15. For the example in Plate 73, s_D^2 = 1.82. This is an estimate of the population
    _____ of _____ scores.

■ ■ ■ ■ ■ ■ ■ ■ ■ ■ ■ ■ ■ ■

variance, difference

16. Next, in order to obtain an estimate of the standard error of the mean
    difference scores, use Formula 18a.

Formula 18a. Estimation of the Population Standard Error of
the Mean Difference Scores

$$s_{\bar{D}} = \sqrt{\frac{s_D^2}{N}}$$

In our example, $s_{\bar{D}} = \underline{\hspace{1cm}}$ (to three decimal places)

■ ■ ■ ■ ■ ■ ■ ■ ■ ■ ■ ■ ■ ■

.427

Formula 18b.

$$s_{\bar{D}} = \sqrt{\frac{N\Sigma D^2 - (\Sigma D)^2}{N^2(N-1)}}$$

17. Formula 18b is algebraically equivalent to Formula 18a and can be used in
    cases where you do not wish to compute the value of _____.

■ ■ ■ ■ ■ ■ ■ ■ ■ ■ ■ ■ ■ ■

$$s_D^2$$

18. Thus, the estimate of the standard error of the mean difference scores in Plate 73 is $s_{\bar{D}} = .427$. In this example, $\bar{X}_1 = \underline{\hspace{1cm}}$ and $\bar{X}_2 = \underline{\hspace{1cm}}$. The mean difference score is $\bar{X}_1 - \bar{X}_2 = \underline{\hspace{0.5cm}}$.

■ ■ ■ ■ ■ ■ ■ ■ ■ ■ ■ ■ ■ ■

3.1, 2.5, .6

Formula 19. Calculation of the t Ratio for Non-independent Means

$$t = \frac{\bar{X}_1 - \bar{X}_2}{s_{\bar{D}}}$$

Degrees of Freedom: $N - 1$ pairs of scores

19. PLATE 73. Determine the t ratio by using Formula 19.

$t = \underline{\hspace{4cm}}$ (to three decimal places)

■ ■ ■ ■ ■ ■ ■ ■ ■ ■ ■ ■ ■ ■

$$\frac{.6}{.427} = 1.405$$

---

20. PLATE 73. $t = 1.405$, $N = 10$. Formula 19 indicates that the degree of freedom associated with this t test are:

df = number of pairs of scores $- 1 = \underline{\hspace{0.5cm}}$.

■ ■ ■ ■ ■ ■ ■ ■ ■ ■ ■ ■ ■ ■

9

---

21. PLATE 73. $t = 1.405$, $df = 9$. Use Table 2 to determine the value of t needed for significance with $df = 9$. For $P = .05$, $t = \underline{\hspace{2cm}}$; for $P = .01$, $t = \underline{\hspace{2cm}}$.

■ ■ ■ ■ ■ ■ ■ ■ ■ ■ ■ ■ ■ ■

2.262, 3.250

---

22. PLATE 73. $t = 1.405$, $df = 9$. From Table 2, the value of the t ratio needed for significance at the .05 level is 2.262. The obtained t ratio of 1.405 is not as large as that required for significance; therefore, you conclude that the difference between the means for the first testing and the second testing is _____ (significant/non-significant).

■ ■ ■ ■ ■ ■ ■ ■ ■ ■ ■ ■ ■ ■

non-significant

23. In making a t test, if the two sets of scores are for the same group of individuals, or if the individuals in the two groups have been paired by some matching technique, the two sets of scores are said to be _____ _____ (independent/non-independent).

■ ■ ■ ■ ■ ■ ■ ■ ■ ■ ■ ■ ■ ■

non-independent

24. In making a t test, if the two sets of scores are for different groups of individuals who have not been paired or matched in any manner, the two sets of scores are said to be _____.

■ ■ ■ ■ ■ ■ ■ ■ ■ ■ ■ ■ ■ ■

independent

25. To perform a t test between independent means you use Formula 16. For non-independent means you use Formula 19. Notice that the numerator _____(is/is not) identical in these two formulas; the denominator____ _____(is/is not) identical; the degrees of freedom_____(are/are not) identical.

■ ■ ■ ■ ■ ■ ■ ■ ■ ■ ■ ■ ■ ■

is, is not, are not

1. What is meant by a non-independent set of scores?

2. A group of sixth-grade students were given Form A and Form B of an intelligence test. The following statistics were obtained. Compute the t ratio, using Formulas 17, 18a, and 19.

Child	Form A	Form B
1	9	10
2	12	14
3	15	13
4	10	7
5	11	7
6	15	12
7	8	9
8	9	12
9	7	6
10	10	14

3. In Exercise 2, assume the researcher did not predict that the scores on one form of the test would be higher than on the other form. Using Table 2, would you reject the null hypothesis that there was no difference between Form A and Form B? Use $P = .05$ as your designated level for significance.

4. For one year, two groups of children were instructed in art by different methods. Each child receiving Method A was matched by IQ score with a child receiving Method B. At the end of the year, each sample of children was given a creativity test. The scores are presented below. Calculate the t ratio using Formulas 18b and 19.

Pairs of Children	Method A	Method B
1st	42	36
2nd	39	38
3rd	37	32
4th	37	31
5th	34	25
6th	32	28
7th	31	21
8th	27	20

5. The researcher in Exercise 4 did not predict which method would be more effective in teaching creativity. Using Table 2, would you

reject the null hypothesis that there is no difference between the two methods of teaching art? Use $P = .05$ as your designated level for significance.

6. Identify the symbol D.

ONE- AND TWO-TAILED TESTS /
TYPE I AND TYPE II ERRORS

Sometimes a researcher makes a hypothesis which states that one of his experimental groups will show the greatest increase. At other times he is not willing to hypothesize which group will change the most. The type of research hypothesis he makes determines how he can evaluate the significance of the difference he obtains. In this set the difference between a one- and a two-tailed test of significance will be explained, as well as how they are conducted.

Because of the sampling error involved in using sample data, a researcher can never be absolutely certain that he has made the right choice in accepting or rejecting the null hypothesis. If he accepts the null hypothesis, there is always some measure of probability that he has been wrong in his decision. If he rejects it, there is also some probability that this decision is in error. These two types of possible errors and their probability of occurrence are discussed in this set.

SPECIFIC OBJECTIVES OF SET 17

At the conclusion of this set you will be able to:

(1) determine which research hypotheses require one-tailed and which require two-tailed tests of significance.

(2) use Table 2 to determine probability levels for one- and two-tailed tests of significance.

(3) state what is meant by a Type I error.

(4) state what is meant by a Type II error.

(5) state how the significance level is related to Type I and Type II errors.

1. PLATE 72, page 203. Notice that the probability of $P = .05$ is divided, with the left "tail" of the distribution containing $P =$ _____ and the right "tail" of the distribution containing $P =$ _____ .

■ ■ ■ ■ ■ ■ ■ ■ ■ ■ ■ ■ ■ ■

.025, .025

2. PLATE 72. Because you are concerned with the probabilities of the two "tails" of the distribution, this is called a _____ -tailed test of significance.

■ ■ ■ ■ ■ ■ ■ ■ ■ ■ ■ ■ ■ ■

two

3. A two-tailed test of significance is concerned with the probability that the obtained $\bar{X}$ difference score lies in _____ (one/either) tail of the distribution.

■ ■ ■ ■ ■ ■ ■ ■ ■ ■ ■ ■ ■ ■

either

4. If a "two-tailed" test of significance is concerned with the probability that an obtained $\bar{X}$ difference score lies in either tail of the distribution, it follows that a "one-tailed" test of significance is concerned with the probability that an obtained $\bar{X}$ difference score lies in _____ tail of the distribution.

■ ■ ■ ■ ■ ■ ■ ■ ■ ■ ■ ■ ■ ■

one

5. A discussion of the types of hypotheses is needed. Recall that the null hypothesis states that there is _____ difference between two groups.

■ ■ ■ ■ ■ ■ ■ ■ ■ ■ ■ ■ ■ ■

no

6. The hypothesis of concern to the researcher may state that there is a group difference. This is usually termed an "alternate" hypothesis. Thus, a hypothesis that states "Tenth-grade boys and tenth-grade girls differ in their heights" is an _____ hypothesis.

■ ■ ■ ■ ■ ■ ■ ■ ■ ■ ■ ■ ■

alternate

7. The alternate hypothesis "Tenth-grade boys and tenth-grade girls differ in their heights" does not imply the *direction* of the difference; that is, whether boys are taller than girls or vice versa. It is only concerned with whether there is a difference in either direction. To "test" this hypothesis, you would use a _____(one/two) tailed test.

■ ■ ■ ■ ■ ■ ■ ■ ■ ■ ■ ■ ■

two

8. If a hypothesis does not hypothesize the direction of the difference (that is, whether one specific group is higher than another), the appropriate test of significance is the two-tailed test. If it does hypothesize the direction of the difference, the appropriate test of significance is the _____-tailed test.

■ ■ ■ ■ ■ ■ ■ ■ ■ ■ ■ ■ ■

one

9. The hypothesis "Intelligence test scores of urban children differ from intelligence test scores of rural children" requires a _____-tailed test of significance.

■ ■ ■ ■ ■ ■ ■ ■ ■ ■ ■ ■ ■

two

10. The hypothesis "Intelligence test scores of urban children differ from intelligence test scores of rural children" requires a two-tailed test of significance because it does not hypothesize the _____of the difference.

■ ■ ■ ■ ■ ■ ■ ■ ■ ■ ■ ■ ■

direction

11. The hypothesis "Intelligence test scores of urban children are higher than the intelligence test scores of rural children" requires a _____-tailed test of significance.

■ ■ ■ ■ ■ ■ ■ ■ ■ ■ ■ ■ ■ ■

one

---

12. The hypothesis "Intelligence test scores of urban children are higher than the intelligence test scores of rural children" requires a one-tailed test of significance because the _____ of the difference is hypothesized.

■ ■ ■ ■ ■ ■ ■ ■ ■ ■ ■ ■ ■ ■

direction

---

13. PLATE 71, page 197. The test of significance that was made for these data was a _____-tailed test because the direction of the difference_____ _____ (was/was not) hypothesized.

■ ■ ■ ■ ■ ■ ■ ■ ■ ■ ■ ■ ■ ■

two, was not

---

14. PLATE 71. If, before the collection of the data, the researcher had made the hypothesis "Tenth-grade boys are taller than tenth-grade girls," he could have made a _____-tailed test of significance.

■ ■ ■ ■ ■ ■ ■ ■ ■ ■ ■ ■ ■ ■

one

---

15. In making a one-tailed test, you perform the t test in exactly the same manner as with the two-tailed test. However, in evaluating the significance of the t ratio for a one-tailed test, you use only _____ tail of the t distribution in determining the probability.

■ ■ ■ ■ ■ ■ ■ ■ ■ ■ ■ ■ ■ ■

one

Plate 74. *t* Distribution for *df* = 61 (One-Tailed Test)

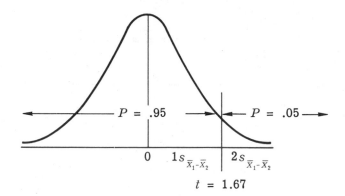

$P = .95$ ← → $P = .05$ →

$0 \quad 1s_{\bar{X}_1 - \bar{X}_2} \quad 2s_{\bar{X}_1 - \bar{X}_2}$

$t = 1.67$

16. This plate presents the *t* distribution for *df* = 61. For the one-tailed test, notice that the probability level (*P* = .05) is indicated on _____(one/two) tail(s) of the distribution.

■ ■ ■ ■ ■ ■ ■ ■ ■ ■ ■ ■ ■

one

17. PLATE 74. The area of the curve that designates the .05 level of significance in a one-tailed test is designated at the right tail of the curve. Recall, from Plate 72, that for a two-tailed test, the .05 level of significance is divided between the _____ tails of the distribution.

■ ■ ■ ■ ■ ■ ■ ■ ■ ■ ■ ■ ■

two

18. TABLE 2. This table presents *t* values that designate probability levels for both tails of the distribution. Therefore, these probability levels, as listed at the top of the table (*P* = .10, *P* = .05, *P* = .02, *P* = .01), are for _____ -tailed tests.

■ ■ ■ ■ ■ ■ ■ ■ ■ ■ ■ ■ ■

two

19. TABLE 2. The probability levels listed are for two-tailed tests. To determine the probability levels appropriate for one-tailed tests, you must divide the listed P values by 2. Thus, the t values for $P = .05$ for a one-tailed test appear in the column headed $P =$ _____ .

■ ■ ■ ■ ■ ■ ■ ■ ■ ■ ■ ■ ■ ■ ■

.10

---

20. TABLE 2. For a one-tailed test, the P values of this table must be divided by 2. Therefore, for a one-tailed test, the heading of the columns should be: $P = .05, P = .025, P =$ _____ , and $P =$ _____ .

■ ■ ■ ■ ■ ■ ■ ■ ■ ■ ■ ■ ■ ■ ■

.01, .005

---

21. To determine the t ratio required at the .05 level of significance for a two-tailed test, use the column headed $P = .05$. To determine the t ratio required at the .05 level of significance for a one-tailed test, use the column headed $P =$ _____ .

■ ■ ■ ■ ■ ■ ■ ■ ■ ■ ■ ■ ■ ■ ■

.10

---

22. PLATE 71, page 197. TABLE 2. $df = 61$.
The t-ratio required for significance at the .05 level, for a two-tailed test, is 2.000 (in column $P = .05$). The t ratio required for significance at the .05 level for a one-tailed test is _____ (in column $P =$ _____).

■ ■ ■ ■ ■ ■ ■ ■ ■ ■ ■ ■ ■ ■ ■

1.671, .10

---

23. PLATE 71. $df = 61$.
For a two-tailed test, the t value required for $P = .05$ is 2.000. For a one-tailed test, the t value required for $P = .05$ is 1.671. Thus, the required value of t for $P = .05$ is _____ (smaller/larger) for a one-tailed test than for a two-tailed test.

■ ■ ■ ■ ■ ■ ■ ■ ■ ■ ■ ■ ■ ■ ■

smaller

---

24. Recall that you can only make a one-tailed test of significance when your alternate hypothesis states the direction of the difference. When there is no direction hypothesized, you must use a _____ -tailed test.

■ ■ ■ ■ ■ ■ ■ ■ ■ ■ ■ ■ ■ ■

two

---

25. TABLE 2. For df = 21, the t ratio required for significance at P = .05 for a two-tailed test is _____ . The t ratio required for a one-tailed test is _____ .

■ ■ ■ ■ ■ ■ ■ ■ ■ ■ ■ ■ ■ ■

2.080, 1.721

---

26. TABLE 2. For df = 30, the t ratio required for significance at P = .01 for a two-tailed test is _____ . The t ratio required for a one-tailed test is _____ .

■ ■ ■ ■ ■ ■ ■ ■ ■ ■ ■ ■ ■ ■

2.750, 2.457 (from column P = .02)

---

27. Recall that the t ratio is obtained in exactly the same manner (Formulas 16 or 19) for a one-tailed test or a two-tailed test. The difference between the two types of tests is in the required value of ___(symbol) for significance at the different probability levels.

■ ■ ■ ■ ■ ■ ■ ■ ■ ■ ■ ■ ■ ■

t

---

28. The hypothesis "Rats that are fed a high protein diet will grow faster than rats that are not fed a high protein diet" requires a _____ -tailed test of significance.

■ ■ ■ ■ ■ ■ ■ ■ ■ ■ ■ ■ ■ ■

one

---

29. The hypothesis "There will be a difference in the arithmetic scores of fifth-grade children who are taught by method A and those who are taught by method B" permits a _____-tailed test of significance.

■ ■ ■ ■ ■ ■ ■ ■ ■ ■ ■ ■ ■ ■

two

---

30. On the basis of the statistical test, you either accept or _____ the null hypothesis.

■ ■ ■ ■ ■ ■ ■ ■ ■ ■ ■ ■ ■ ■

reject

---

31. Recall that there are two types of hypotheses: the statistical hypothesis, which is called the _____ hypothesis, and the alternate hypothesis.

■ ■ ■ ■ ■ ■ ■ ■ ■ ■ ■ ■ ■ ■

null

---

32. The hypothesis that may state there is a difference between two groups is called a(n)_____ hypothesis.

■ ■ ■ ■ ■ ■ ■ ■ ■ ■ ■ ■ ■ ■

alternate

---

33. In addition to hypothesizing a difference between two groups, the alternate hypothesis may also state the _____ of the difference.

■ ■ ■ ■ ■ ■ ■ ■ ■ ■ ■ ■ ■ ■

direction

---

34. The statistical hypothesis "tested" by statistical methods is the null hypothesis, which states there is ____ difference between the groups.

■ ■ ■ ■ ■ ■ ■ ■ ■ ■ ■ ■ ■ ■

no

---

35. When the alternate hypothesis does not state the direction of the difference, you use a _____ (one/two) tailed test.

■ ■ ■ ■ ■ ■ ■ ■ ■ ■ ■ ■ ■ ■

two

36. When the alternate hypothesis does state the direction of the difference, you use a _____ (one/two) tailed test.

■ ■ ■ ■ ■ ■ ■ ■ ■ ■ ■ ■ ■ ■

one

37. For either type of alternate hypothesis, the statistical hypothesis to be "tested" is the _____ hypothesis.

■ ■ ■ ■ ■ ■ ■ ■ ■ ■ ■ ■ ■ ■

null

38. When using sample data, you *never* are able to accept or reject the null hypothesis with absolute certainty, because there is always the possibility that you are wrong, due to _____ error.

■ ■ ■ ■ ■ ■ ■ ■ ■ ■ ■ ■ ■ ■

sampling

39. However, on the basis of a *t* test using sample data, you make the decision either to accept or reject the null hypothesis. Researchers generally make the decision to reject the null hypothesis if the probability that their decision is incorrect is as small or smaller than $P =$ _____ .

■ ■ ■ ■ ■ ■ ■ ■ ■ ■ ■ ■ ■ ■

.05

40. The use of the probability of $P = .05$ in making the decision to accept or _____ the null hypothesis is purely an arbitrary one, but it is the level of significance usually accepted by most researchers.

■ ■ ■ ■ ■ ■ ■ ■ ■ ■ ■ ■ ■ ■ ■

reject

Plate 75.

Type I Error: Rejecting the null hypothesis on the basis of sample data, when, in fact, no difference exists.

Type II Error: Accepting the null hypothesis on the basis of sample data, when, in fact, a true difference exists.

41. You can make two types of error when you decide either to accept or to reject the null hypothesis. Read the statements in Plate 75. The two types of error are called _____ error and _____ error.

■ ■ ■ ■ ■ ■ ■ ■ ■ ■ ■ ■ ■ ■ ■

Type I, Type II

42. PLATE 75. If, on the basis of your statistical test, you have accepted the null hypothesis when, in fact, the two samples were selected from different populations of scores, you have made a Type ___(I/II) error.

■ ■ ■ ■ ■ ■ ■ ■ ■ ■ ■ ■ ■ ■ ■

II

43. PLATE 75. If, on the basis of your statistical test, you have rejected the null hypothesis when, in fact, the two samples were selected from the same population of scores, you have made a Type ___(I/II) error.

■ ■ ■ ■ ■ ■ ■ ■ ■ ■ ■ ■ ■ ■ ■

I

44. In rejecting a null hypothesis on the basis of sample data, there is always some probability that you have made a Type I error, because the difference that you find between samples may be due to _____ error.

■ ■ ■ ■ ■ ■ ■ ■ ■ ■ ■ ■ ■ ■

sampling

45. If you reject the null hypothesis at the .05 level of significance, the probability that you have made a Type I error is only $P =$ _____ .

■ ■ ■ ■ ■ ■ ■ ■ ■ ■ ■ ■ ■ ■

.05

46. If you reject the null hypothesis at the .01 level of significance, the probability that you have made a Type I error is only $P =$ _____ .

■ ■ ■ ■ ■ ■ ■ ■ ■ ■ ■ ■ ■ ■

.01

47. The smaller the level of significance, the _____ (smaller/greater) the probability that you have made a Type I error.

■ ■ ■ ■ ■ ■ ■ ■ ■ ■ ■ ■ ■ ■

smaller

EXERCISES

1. What is the difference between a one-tailed and a two-tailed test of significance?

2. Should a one-tailed or a two-tailed test of significance be employed when testing the following hypotheses?
   (a) Students exposed to kinesthetic reading methods will show greater gains on the California Primary Reading Achievement Test than will a comparable group of students exposed to phonic reading methods.
   (b) Students involved in intensive counseling will show greater gains in the Pilgrim Adjustment Inventory than will comparable students not receiving such counseling.
   (c) There will be a difference in the school drop-out rate for students of differing social classes.
   (d) Students who take a traditional social studies course will show a different rate of progress in cognitive development than comparable students taking the Smith Social Studies Curriculum.

3. Using Table 2, determine the t ratio required for a two-tailed test of significance for $P = .05$ when $df = 12$. Determine the t ratio required for a one-tailed test.

4. Using Table 2, determine the t ratio required for a two-tailed test of significance for $P = .01$ when $df = 23$. Determine the t ratio required for a one-tailed test.

5. For the same df, does a one-tailed test or a two-tailed test require a larger t ratio?

6. What is meant by a Type I error?

7. What is meant by a Type II error?

8. How is the size of the level of significance related to the probability that, in rejecting the null hypothesis, you have made a Type II error? A Type I error?

ANALYSIS OF VARIANCE

The t test has provided a method for analyzing the difference between two sample means. Occasionally a research design requires that statistical techniques be applied to the data contained in three or more samples.

A common statistical technique which permits an analysis of the data in more than two samples at a time is called the *analysis-of-variance* technique. As implied by the name of this technique, it is concerned with analyzing the variance estimates obtained from the various samples. This set will indicate how the total variance contained in a number of samples can be divided into component parts and analyzed in order to determine the significance of the difference between the samples. The form of analysis of variance presented in this set and in set 19 is the simplest form involving three or more samples.

SPECIFIC OBJECTIVES OF SET 18

At the conclusion of this set you will be able to:

(1) state the two components into which the total sum of squares are divided in the analysis-of-variance technique.

(2) use formulas 22 through 24 to compute the sums of squares for the components in the analysis-of-variance test for the significance of differences between three or more samples.

(3) use formulas 25 through 27 to compute the degrees of freedom associated with the components in the analysis-of-variance technique.

(4) identify the symbols SS_b, SS_w, SS_t, MS.

1. The *t* test, you will recall, is used to test the null hypothesis that there is no difference between the mean scores of _____ populations.

■ ■ ■ ■ ■ ■ ■ ■ ■ ■ ■ ■ ■ ■

two

---

2. In research designs involving more than two samples you may also wish to test the null hypothesis. If you have three samples, for instance, the null hypothesis would state that there was no difference between the mean scores of _____ populations.

■ ■ ■ ■ ■ ■ ■ ■ ■ ■ ■ ■ ■ ■

three

---

3. Since the *t* test is useful only for testing the difference between two means, you must apply another statistical technique, called *analysis of variance*, in cases in which there are _____ or more samples.

■ ■ ■ ■ ■ ■ ■ ■ ■ ■ ■ ■ ■

three

Plate 76. An Experiment in Teaching Arithmetic

The Centerville school wished to determine if certain methods of teaching arithmetic were better than others. It devised three different methods of teaching arithmetic and divided twenty-one children into three groups, giving each group a different method. The three groups were designated Group A, Group B, and Group C. These three groups of children did not differ initially on arithmetic achievement. At the end of the school year an arithmetic achievement test was given to each of the children in the three groups, and the following scores were obtained:

Group A	Group B	Group C
94	90	86
92	86	84
90	84	80
90	82	78
86	82	78
86	80	72
84	78	
82		
$N_A = 8$	$N_B = 7$	$N_C = 6$
$\bar{X}_A = 88$	$\bar{X}_B = 83.143$	$\bar{X}_C = 79.667$

4. The experiment presented in Plate 76 is concerned with examining the difference between _____ methods of teaching arithmetic.

■ ■ ■ ■ ■ ■ ■ ■ ■ ■ ■ ■ ■ ■

three

---

5. The null hypothesis being tested in this experiment is that there is no _____ between the three methods of teaching arithmetic.

■ ■ ■ ■ ■ ■ ■ ■ ■ ■ ■ ■ ■ ■

difference

---

6. Notice that in Plate 76 there is a mean score reported for each group. Observe that the $\bar{X}$'s of the three groups ____(do/do not) differ.

■ ■ ■ ■ ■ ■ ■ ■ ■ ■ ■ ■ ■ ■

do

---

7. Since the means of the three samples differ, your concern here is determining whether these differences are large enough to represent real differences or whether they are due to _____ error.

■ ■ ■ ■ ■ ■ ■ ■ ■ ■ ■ ■ ■ ■

sampling

---

8. When more than two samples are involved you can decide whether or not to accept the null hypothesis by using the *analysis of*_____ technique.

■ ■ ■ ■ ■ ■ ■ ■ ■ ■ ■ ■ ■ ■

variance

---

9. Plate 76 indicates that the raw scores within each of the three groups____ (do/do not) differ.

■ ■ ■ ■ ■ ■ ■ ■ ■ ■ ■ ■ ■ ■

do

---

10. Plate 76 also indicates that the means of the three groups ____ (do/do not) differ.

■ ■ ■ ■ ■ ■ ■ ■ ■ ■ ■ ■ ■ ■

do

---

11. Comparing the variance among the means of the several groups with the variance among the scores within the groups is the major focus of the analysis-of- _____ technique.

■ ■ ■ ■ ■ ■ ■ ■ ■ ■ ■ ■ ■ ■

variance

Plate 77. Analysis of Variance for Experiment Presented in Plate 76

	Sum of Squares	Degrees of Freedom	Mean Square (Variance Estimate)	F
Between groups	$SS_b = 245.8$	$df_b = 2$	$MS_b = 122.9$	6.54
Within groups	$SS_w = 338.2$	$df_w = 18$	$MS_w = 18.8$	
Total	$SS_t = 584$	$df_t = 20$		

12. Plate 77 presents the usual method of reporting analysis-of-variance tests. We shall examine this plate, which presents the statistical test of the data in Plate 76, and then we shall learn how it was computed.

■ ■ ■ ■ ■ ■ ■ ■ ■ ■ ■ ■ ■ ■

go on to next frame

---

13. PLATE 77. The new symbols introduced in this plate can be readily learned. SS means sum of squares, MS means _____ _____ , and the subscripts b, w, and t indicate _____ , _____ , and _____ .

■ ■ ■ ■ ■ ■ ■ ■ ■ ■ ■ ■ ■ ■

mean square, between, within, total

14. PLATE 77. As indicated in the third column, the variance estimate is indicated by a new term called _____ _____ .

■ ■ ■ ■ ■ ■ ■ ■ ■ ■ ■ ■ ■ ■

mean square

---

15. The term *mean square* is used to indicate estimates of variance in the analysis-of-variance technique. You will notice that two mean squares have been computed: the mean square _____ groups, and the mean square _____ groups.

■ ■ ■ ■ ■ ■ ■ ■ ■ ■ ■ ■ ■ ■

between, within

---

16. The analysis-of-variance technique tests the difference between these two mean squares.

Formula 20. *F* Test. Formula for Computing the *F* Ratio in the Analysis of Variance
$$F = \frac{\text{Mean Square between groups}}{\text{Mean Square within groups}} = \frac{MS_b}{MS_w}$$

Formula 20 presents the formula for the analysis-of-variance test of significance, commonly known as the __(symbol) test.

■ ■ ■ ■ ■ ■ ■ ■ ■ ■ ■ ■ ■ ■

F

---

17. Formula 20 indicates that in order to determine the value of F you must divide the mean square _____ groups by the mean square _____ groups.

■ ■ ■ ■ ■ ■ ■ ■ ■ ■ ■ ■ ■ ■

between, within

---

18. In Plate 77, the F ratio of 6.54 has been determined by applying Formula 20. Thus, in this example, F was obtained by dividing _____ by_____.

■ ■ ■ ■ ■ ■ ■ ■ ■ ■ ■ ■ ■ ■

122.9, 18.8

---

19. We shall return later to determine the significance level of this F ratio. Now let us determine how we obtained these mean squares. Examine the first column in Plate 77, which presents three _____ __ _____ .

■ ■ ■ ■ ■ ■ ■ ■ ■ ■ ■ ■ ■ ■

sums of squares

---

20. PLATE 77. The first column reveals that the total sum of squares can be broken down into two parts: the sum of squares _____ groups, and the sum of squares _____ groups.

■ ■ ■ ■ ■ ■ ■ ■ ■ ■ ■ ■ ■ ■

between, within

---

21. The fact that the sum of squares between the groups (SS_b) plus the sum of squares within the groups (SS_w) equals the _____ sum of squares is the basis for the analysis-of-variance technique.

■ ■ ■ ■ ■ ■ ■ ■ ■ ■ ■ ■ ■ ■

total

---

22. To understand this, let us examine the frequency distributions presented in Plate 78. The first column presents the frequency distribution of arithmetic scores for all three groups combined. The next three columns present the frequency distribution for Group ___, Group ___, and Group ___, respectively.

Plate 78. Frequency Distributions of Arithmetic Scores for Groups A, B, and C and for All Three Combined

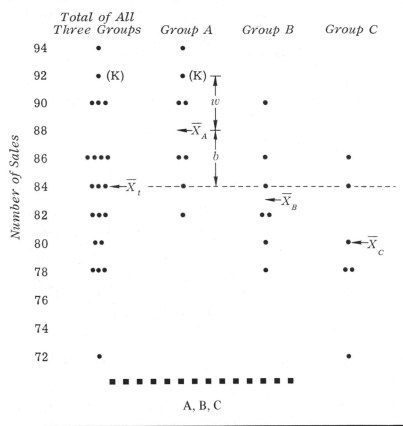

23. PLATE 78. As can be seen, the mean score for the total distribution is 84, and is designated by the symbol _____.

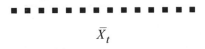

$$\bar{X}_t$$

24. PLATE 78. Notice that the mean scores for the group distributions are also indicated at points $\bar{X}_A$, $\bar{X}_B$, and $\bar{X}_C$. The level of $\bar{X}_t$ is indicated by the horizontal dotted line. It is evident from this plate that the means of the groups _____ (do/do not) differ from $\bar{X}_t$.

■ ■ ■ ■ ■ ■ ■ ■ ■ ■ ■ ■ ■

do

25. Recall that it was stated that the sum of squares for the total (SS_t) can be broken down into the sum of squares between groups (SS_b) and sum of squares within groups (SS_w). This can be illustrated by examining one of the scores in Plate 78. For example, take Subject K in Group A. Subject K has a raw score of ____ .

■ ■ ■ ■ ■ ■ ■ ■ ■ ■ ■ ■ ■ ■

92

26. PLATE 78. Subject K's score of 92 deviates from $\bar{X}_t$ by ____ points.

■ ■ ■ ■ ■ ■ ■ ■ ■ ■ ■ ■ ■ ■

8

27. Subject K's deviation of 8 points from $\bar{X}_t$ can be divided into two portions:
   (1) Subject K's score deviates from its own Group mean $(\bar{X}_A)$ by ____ points.
   (2) The mean of Group A $(\bar{X}_A)$ deviates from the total mean (X_t) by ____ points.

■ ■ ■ ■ ■ ■ ■ ■ ■ ■ ■ ■ ■ ■

4, 4

28. Thus, the deviation of subject K's score from the total mean $(\bar{X}_t)$ is exactly equal to the sum of its deviation from its group mean $(\bar{X}_A)$ and the deviation of ____ (symbol) from $\bar{X}_t$.

■ ■ ■ ■ ■ ■ ■ ■ ■ ■ ■ ■ ■ ■

$\bar{X}_A$

29. PLATE 78. It has been shown that the deviation of any one score in Group A from $\bar{X}_t$ can be divided into two portions: (1) its deviation from ____ (symbol) and (2) the deviation of ____ (symbol) from ____ (symbol). This is true of the scores in all three groups.

■ ■ ■ ■ ■ ■ ■ ■ ■ ■ ■ ■ ■ ■

$\bar{X}_A , \bar{X}_A , \bar{X}_t$

30. FORMULA 21. By the same token the sum of squared deviations (SS_t) of all the scores in the various groups can be divided into two portions: (1) the sum of squares within the groups (SS_w), and (2) the sum of squares _____ the groups (_____) (symbol).

■ ■ ■ ■ ■ ■ ■ ■ ■ ■ ■ ■ ■ ■

between, (SS_b)

---

31. The value of these sums of squares can be computed by using Formulas 22 through 24. (It is instructive to carry out the calculations indicated by Formulas 22 through 24 directly, although far simpler methods for calculating these quantities will be presented later.) Formula 22 presents the procedure for calculating the total sum of squares. You will recognize that this formula is identical with Formula 8. In our experiment in Plate 76, $N_t = $ _____ .

■ ■ ■ ■ ■ ■ ■ ■ ■ ■ ■ ■ ■ ■

21

---

32. Compute the total sum of squares (SS_t) for the data presented in Plate 76, page 229. Use Formula 22. $SS_t = $ _____

■ ■ ■ ■ ■ ■ ■ ■ ■ ■ ■ ■ ■ ■

584

---

33. PLATE 76. $SS_t = 584$.

As indicated in Formula 23, you need first to calculate the sum of squares within each group separately and then _____ them to obtain SS_w.

■ ■ ■ ■ ■ ■ ■ ■ ■ ■ ■ ■ ■ ■

sum

34. When computing the sum of squares within a group, you apply the same formula (Formula 22) to the data within each group. For instance, in Group A, SS_w is computed using only the raw scores in Group A and $N_A = $ ___ .

■ ■ ■ ■ ■ ■ ■ ■ ■ ■ ■ ■ ■ ■

8

35. PLATE 76. Use Formula 22 to compute the within-group sum of squares for each group. $SS_A = $ ___ ; $SS_B = $ ___ ; $SS_C = $ ___

■ ■ ■ ■ ■ ■ ■ ■ ■ ■ ■ ■ ■ ■

120, 94.9, 123.3

36. PLATE 76. $SS_A = 120, SS_B = 94.9, SS_C = 123.3$.

Formula 23 presents the formula for obtaining the sum of squares within groups. For the data in Plate 76, $SS_w = $ _____ .

■ ■ ■ ■ ■ ■ ■ ■ ■ ■ ■ ■ ■ ■

338.2

37. PLATE 76. You now have computed the total sum of squares (SS_t) and the within-groups sum of squares (SS_w). Formula 24 presents two methods of computing the sum of squares between groups (SS_b). Formula 24a is the easy way because, using it, you merely subtract _____ (symbol) from _____ (symbol).

■ ■ ■ ■ ■ ■ ■ ■ ■ ■ ■ ■ ■ ■

SS_w, SS_t

237

38. Formula 24b presents the computational formula for SS_b. Apply this formula to the data in Plate 76. This formula indicates that you square the deviation of each group mean from the _____ mean. Next, for each group, you multiply this squared deviation by ____ (symbol) of the group. Finally, you sum the products to obtain _____ (symbol).

■ ■ ■ ■ ■ ■ ■ ■ ■ ■ ■ ■ ■ ■

total, N, SS_b

---

39. PLATE 76. Compute SS_b by Formula 24b. SS_b = _____

_____ .

■ ■ ■ ■ ■ ■ ■ ■ ■ ■ ■ ■ ■ ■

$8(88 - 84)^2 + 7(83.143 - 84)^2 + 6(79.667 - 84)^2 = 245.8$

---

40. PLATE 76. $SS_t = 584$, $SS_w = 338.2$.
By Formula 24b, you know that $SS_b = 245.8$. To check the accuracy of your calculations, you apply Formula 24a and find that SS_b = _____ .

■ ■ ■ ■ ■ ■ ■ ■ ■ ■ ■ ■ ■ ■

245.8

---

41. PLATE 76. You have thus shown that SS_w plus SS_b equals _____ (symbol).

■ ■ ■ ■ ■ ■ ■ ■ ■ ■ ■ ■ ■ ■

SS_t

Formulas 25 through 27. Calculation of Degrees of Freedom for Analysis of Variance

	Degrees of freedom	
Total	$df_t = N - 1$	(Formula 25)
Between groups	df_b = No. of groups − 1	(Formula 26)
Within groups	$df_w = df_t - df_b$	(Formula 27)

238

Formulas 28 and 29. Calculation of Mean Squares (variance estimates)

$$MS_b = \frac{SS_b}{df_b} \qquad \text{(Formula 28)}$$

$$MS_w = \frac{SS_w}{df_w} \qquad \text{(Formula 29)}$$

42. PLATE 77. This plate shows the three sums of squares you have just calculated. Formulas 28 and 29 indicate that, in order to compute the mean squares from the sums of squares, you must first determine the _____ __ _____ associated with each *SS*.

■ ■ ■ ■ ■ ■ ■ ■ ■ ■ ■ ■ ■ ■

degrees of freedom

---

43. Formulas 25 through 27 are used in determining the degrees of freedom associated with each sum of squares. The same reasoning applies with these *df* calculations as with the *df* calculations for the *t* test. For example, the degrees of freedom for the total group (df_t) is equal to the total ____ -1.

■ ■ ■ ■ ■ ■ ■ ■ ■ ■ ■ ■ ■ ■

N

---

44. FORMULA 26. Since this formula is concerned with the sum of squares between several groups, the degrees of freedom between groups (df_b) is equal to the _____ of _____ minus one.

■ ■ ■ ■ ■ ■ ■ ■ ■ ■ ■ ■ ■ ■

number, groups

---

45. FORMULA 27. Because the total degrees of freedom (df_t) must equal the sum of df_b and df_w, it follows that df_w can be obtained by subtracting _____ (symbol) from _____ (symbol).

■ ■ ■ ■ ■ ■ ■ ■ ■ ■ ■ ■ ■ ■

df_b, df_t

---

46. Applying Formulas 25 through 27 to the experimental data in Plate 77, you find that when there are three groups, the $df_b = $ ___ .

■ ■ ■ ■ ■ ■ ■ ■ ■ ■ ■ ■ ■ ■

2

---

47. You also find that, since the total N of the combined groups is 21, $df_t = $ ___ .

■ ■ ■ ■ ■ ■ ■ ■ ■ ■ ■ ■ ■ ■

20

---

48. Applying Formula 27, you find that $df_w = $ ___ .

■ ■ ■ ■ ■ ■ ■ ■ ■ ■ ■ ■ ■ ■

18

---

49. PLATE 77. You have now determined the sums of squares and the degrees of freedom associated with them. Applying Formulas 28 and 29, you can easily determine the mean square between groups (MS_b) and the mean square within groups (MS_w). $MS_b = $ _____ . $MS_w = $ _____ .

■ ■ ■ ■ ■ ■ ■ ■ ■ ■ ■ ■ ■ ■

122.9, 18.8

---

50. PLATE 77. Having computed the two mean squares necessary in the analysis-of-variance technique, you are now ready to examine the relationship of these two mean squares. Recall that the term *mean square* is identical with _____ .

■ ■ ■ ■ ■ ■ ■ ■ ■ ■ ■ ■ ■ ■

variance

---

51. You have obtained two estimates of the variance of our population of arithmetic scores. One of these is the variance of the scores within the groups, indicated in Plate 77 by _____ (symbol). The other estimate of the population variance is the variance obtained between the several groups, indicated by _____ (symbol).

■ ■ ■ ■ ■ ■ ■ ■ ■ ■ ■ ■ ■ ■

MS_w, MS_b

EXERCISES

1. What are the two components into which the total sums of squares are divided in the analysis-of-variance technique?

2. Thirty sixth-grade students were divided into three socioeconomic groups and were administered a reading comprehension test. The researcher had the hypothesis that there is a difference between reading comprehension test scores for children of differing socioeconomic backgrounds. The raw scores presented below were obtained to test this hypothesis. Use Formulas 22 through 24 to compute the sums of squares needed for the analysis-of-variance technique.

Group A	Group B	Group C
21	20	24
18	20	23
17	19	23
17	18	22
17	17	20
16	17	20
15	14	18
13	14	17
12	9	16
8	7	
7		

3. For the data presented in Exercise 2, determine the degrees of freedom associated with each sum of squares, using Formulas 25 through 27, and compute the necessary mean squares, using Formula 28.

4. What do the symbols SS_b, SS_w, SS_t, and MS represent?

ANALYSIS OF VARIANCE (CONTINUED) /
F TEST FOR TWO VARIANCE ESTIMATES

The ratio between the mean squares between groups (MS_b) and the mean squares within groups (MS_w) is the basis of the analysis-of-variance technique. The evaluation of this F ratio, using the degrees of freedom associated with the mean square estimates, permits the researcher either to accept or to reject the null hypothesis that there is no difference among the various sample means. Of course, if he can accept the null hypothesis, he can state that there is no difference among the various sample means. If the analysis-of-variance technique permits the researcher to reject the null hypothesis, he then concludes that there are significant differences between the various samples. This set presents the method for computing and evaluating the F ratio. Also, in this set, there is another method of using the F ratio to evaluate the difference between two sample variances.

SPECIFIC OBJECTIVES OF SET 19

At the conclusion of this set you will be able to:

(1) identify the symbol F.

(2) use Formula 20 to compute the F ratio in the analysis-of-variance technique.

(3) use Table 3 to evaluate the F ratio for significance.

(4) use Formula 31 to determine the F ratio between two variance estimates, and Table 3 to evaluate it for significance.

1. Recall that the null hypothesis for the experiment in Plate 76 states that there is no difference between the arithmetic scores of Groups A, B, and C. Your calculations have shown that there is variability within the groups as well as _____ the groups.

■ ■ ■ ■ ■ ■ ■ ■ ■ ■ ■ ■ ■ ■

between

---

2. If you can demonstrate that the variability *between* the groups is so much greater than the variability *within* the groups that there is a low probability that the between-group variance could have been a result of sampling _____ , then you may _____ (accept/reject) the null hypothesis.

■ ■ ■ ■ ■ ■ ■ ■ ■ ■ ■ ■ ■ ■

error, reject

---

3. The question, then, is whether the estimate of the population variance obtained *between* the groups (MS_b) is sufficiently greater than the estimate of the population variance obtained *within* the groups (MS_w) to warrant our rejection of the _____ _____ .

■ ■ ■ ■ ■ ■ ■ ■ ■ ■ ■ ■ ■ ■

null hypothesis

---

4. To make a statistical decision regarding the relationship of MS_b and MS_w we apply the F test as given in Formula 20. As seen in this formula, F is expressed as a ratio of MS_b to _____ (symbol).

■ ■ ■ ■ ■ ■ ■ ■ ■ ■ ■ ■ ■ ■

MS_w

---

5. Compute the F ratio of the mean squares obtained in Plate 77. $F = $ _____ .

■ ■ ■ ■ ■ ■ ■ ■ ■ ■ ■ ■ ■ ■

6.54

6. PLATE 77, page 231. $F = 6.54$.
   As with the t test, you must determine the significance level for this F ratio. Unlike the t test, which had only one value for the degrees of freedom, the F test has _____ values for df. These are df_b and _____ (symbol).

   ■ ■ ■ ■ ■ ■ ■ ■ ■ ■ ■ ■ ■ ■

   two, df_w

---

7. To evaluate the significance of an F ratio you use Table 3, which presents the values of F at the .05 and .01 levels of significance. This table indicates that the df associated with the numerator is listed across the top and the df associated with the denominator is listed along the side. Thus, you must enter this table using both df_w and _____ (symbol).

   ■ ■ ■ ■ ■ ■ ■ ■ ■ ■ ■ ■ ■ ■

   df_b

---

8. TABLE 3. The light type in the body of this table indicates the value needed for significance at the .05 level. The boldface type gives the value needed for the _____ level.

   ■ ■ ■ ■ ■ ■ ■ ■ ■ ■ ■ ■ ■ ■

   .01

---

9. PLATE 77. In order to evaluate the significance of the F ratio of 6.54, you must enter Table 3 with df_b = ___ and df_w = ___ .

   ■ ■ ■ ■ ■ ■ ■ ■ ■ ■ ■ ■ ■ ■

   2, 18

---

10. PLATE 77. The numerator in this example is _____ (symbol). The degrees of freedom associated with it is ___ .

    ■ ■ ■ ■ ■ ■ ■ ■ ■ ■ ■ ■ ■ ■

    MS_b, 2

11. PLATE 77. Thus, to determine the significance of our F ratio, you enter Table 3 with $df = 2$ along the top because it is associated with the _____ . Locate this column in Table 3.

■ ■ ■ ■ ■ ■ ■ ■ ■ ■ ■ ■ ■ ■

numerator

12. Likewise, you enter Table 3 with $df = 18$ along the side of the table because it is associated with the _____. Locate this row in Table 3.

■ ■ ■ ■ ■ ■ ■ ■ ■ ■ ■ ■ ■ ■

denominator

13. TABLE 3. At the intersection of the column ($df = 2$) and the row ($df = 18$) you find that the F ratio needed for $P = .05$ is _____ and for $P = .01$ is _____.

■ ■ ■ ■ ■ ■ ■ ■ ■ ■ ■ ■ ■ ■

3.55, 6.01

14. PLATE 77. You have determined that, in order to be significant at the .05 level, the F ratio must be at least 3.55. In order for it to be significant at the .01 level, it must be at least 6.01. Your obtained F ratio is 6.54. This value _____ (is/is not) greater than is required for significance at the .01 level.

■ ■ ■ ■ ■ ■ ■ ■ ■ ■ ■ ■ ■

is

15. Your F ratio of 6.54 exceeds that required at the .01 level of significance. Therefore, you are justified in _____ (accepting/rejecting) the null hypothesis that there are no differences between the three methods of teaching arithmetic.

■ ■ ■ ■ ■ ■ ■ ■ ■ ■ ■ ■ ■ ■

rejecting

16. PLATE 77. The variance between the groups (MS_b) has been shown to be sufficiently greater than the variance within the groups (MS_w). Thus, you can confidently say that the three groups of scores did not come from the same _____ of scores.

■ ■ ■ ■ ■ ■ ■ ■ ■ ■ ■ ■ ■ ■

population

Note: You now know that there is a difference between the three methods of teaching arithmetic. However, you do not know whether there is a significant difference between all three methods, or whether it is just between one method and the other two. The advanced statistical techniques for making these determinations are beyond the scope of this book.

17. In using the analysis-of-variance technique, there is a computationally easier method for obtaining the sums of squares. This method and its format are presented in Formula 30. As shown in the columns, this method requires that first you obtain, for each group and for the total, the values of _____ (symbol), _____ (symbol), and ____ (symbol).

■ ■ ■ ■ ■ ■ ■ ■ ■ ■ ■ ■ ■ ■

$\Sigma X, \Sigma X^2, N$

Formula 30. Computational Formulas for Sums of Squares

	Group A	Group B	Group C	Total
	ΣX_A	ΣX_B	ΣX_C	ΣX_t
	$\Sigma X_A{}^2$	$\Sigma X_B{}^2$	$\Sigma X_C{}^2$	$\Sigma X_t{}^2$
	N_A	N_B	N_C	N_t

Step 1   Correction term $C = \dfrac{(\Sigma X_t)^2}{N_t}$

Step 2   Total sum of squares $SS_t = \Sigma X_t{}^2 - C$

Step 3   Sum of squares between groups $SS_b = \dfrac{(\Sigma X_A)^2}{N_A} + \dfrac{(\Sigma X_B)^2}{N_B} + \dfrac{(\Sigma X_C)^2}{N_C} - C$

Step 4   Sum of squares within groups $SS_w = SS_t - SS_b$

18. **FORMULA 30.** Substituting these values in the algebraic formulas, you first compute, as shown in Step 1, a _____ term to be used in Steps 2 and 3.

■ ■ ■ ■ ■ ■ ■ ■ ■ ■ ■ ■ ■ ■

correction

---

19. **FORMULA 30.** Step 2 gives the formula for computing the total sum of squares (SS_t). Step 3 gives the formula for computing the sum of squares between the groups (SS_b). Step 4 indicates that to obtain the sum of squares within the groups (SS_w) you subtract _____ (symbol) from _____ (symbol).

■ ■ ■ ■ ■ ■ ■ ■ ■ ■ ■ ■ ■ ■

SS_b, SS_t

---

20. Using these formulas, calculate the sums of squares for the data presented in Plate 76. The sums of squares calculated by using Formula 30 should be identical with those previously computed for Plate 77.

■ ■ ■ ■ ■ ■ ■ ■ ■ ■ ■ ■ ■ ■

Plate 79. Computation of Sum of Squares for Arithmetic Scores Presented in Plate 76

Group A	Group B	Group C	
94	90	86	
92	86	84	
90	84	80	
90	82	78	
86	82	78	
86	80	72	
84	78		
82			
			Total:
$\Sigma X_A = 704$	$\Sigma X_B = 582$	$\Sigma X_C = 478$	$\Sigma X_t = 1764$
$\Sigma X_A{}^2 = 62072$	$\Sigma X_B{}^2 = 48484$	$\Sigma X_C{}^2 = 38204$	$\Sigma X_t{}^2 = 148760$
$\bar{X}_A = 88$	$\bar{X}_B = 83.143$	$\bar{X}_C = 79.667$	$\bar{X}_t = 84.0$
$N_A = 8$	$N_B = 7$	$N_C = 6$	$N_t = 21$

248

Step 1	$$C = \frac{(1764)^2}{21} = 148176$$
Step 2	$$SS_t = 148760 - 148176 = 584$$
Step 3	$$SS_b = \frac{(704)^2}{8} + \frac{(582)^2}{7} + \frac{(478)^2}{6} - 148176 = 245.8$$
Step 4	$$SS_w = 584 - 245.8 = 338.2$$

Note: This is the simplest form of the analysis-of-variance technique. It can be applied to any number of groups simply by extending the formulas to include the additional groups. Also, the analysis-of-variance technique can be applied to experiments involving groups within groups. However, these advanced techniques are beyond the scope of this book.

21. Formula 20 indicates that in computing the F ratio, you always use _____ (symbol) as the numerator and _____ (symbol) as the denominator.

■ ■ ■ ■ ■ ■ ■ ■ ■ ■ ■ ■ ■ ■

MS_b, MS_w

22. The F test provides you with the probability that the mean square between groups (MS_b) is _____ (larger/smaller) than the mean square within groups (MS_w). Therefore, the F test in the analysis of variance technique is always a _____ (one/two)-tailed test.

■ ■ ■ ■ ■ ■ ■ ■ ■ ■ ■ ■ ■ ■

larger, one

23. If you reject the null hypothesis on the basis of the F test, you accept the alternate hypothesis that the sample means _____ (are/are not) from the same population.

■ ■ ■ ■ ■ ■ ■ ■ ■ ■ ■ ■ ■ ■

are not

Formula 31. Calculation of the F-Ratio for Comparison of Two Variance Estimates

$$F = \frac{\text{larger } s^2}{\text{smaller } s^2}$$

24. There is a simpler type of F ratio you can compute when you wish only to test the difference between two population variances. Formula 31 indicates that when you have two variance estimates you can obtain the F ratio by dividing the _____ variance estimate by the _____ variance estimate.

■ ■ ■ ■ ■ ■ ■ ■ ■ ■ ■ ■ ■ ■

larger, smaller

Plate 80. Hypothesis: There is a difference in the variability of reading achievement scores between boys and girls.

Boy sample	$s^2 = 15$	$df = 16$
Girl sample	$s^2 = 34$	$df = 12$

25. Plate 80 presents the variance estimates and df for samples of boys and girls on reading achievement scores. The variance estimate obtained in the boy sample is _____ (larger/smaller) than that obtained in the girl sample.

■ ■ ■ ■ ■ ■ ■ ■ ■ ■ ■ ■ ■ ■

smaller

---

26. PLATE 80. The null hypothesis being tested by the F test here is that there is no difference between the two population _____ .

■ ■ ■ ■ ■ ■ ■ ■ ■ ■ ■ ■ ■ ■

variances

---

27. Formula 31 indicates that, to obtain the F ratio, you divide the larger s^2 by the smaller s^2. In our example, you should divide s^2 of the _____ (boys/girls) by the s^2 of the _____ (boys/girls). For Plate 80, $F =$ _____ .

■ ■ ■ ■ ■ ■ ■ ■ ■ ■ ■ ■ ■ ■

girls, boys, 2.27

28. The hypothesis stated in Plate 80 indicates that the investigator is interested in determining if there is a difference in variability, *regardless* of which sex is more variable. However, the F test in Formula 31 requires that the larger s^2 be placed in the numerator. Therefore, if the F ratio exceeds that listed in Table 3 at the .05 level, you can reject the null hypothesis at only the _____ level.

■ ■ ■ ■ ■ ■ ■ ■ ■ ■ ■ ■ ■ ■

.10

29. To determine significant levels for F tests of the type presented in Formula 31, you must double the P values in this table. Thus, for this type of test, values in Roman type are for $P =$ _____, and those in boldface type are for $P =$ _____ .

■ ■ ■ ■ ■ ■ ■ ■ ■ ■ ■ ■ ■ ■

.10, .02

30. PLATE 80. The *df* for the larger variance estimate is $df =$ _____ and for the smaller variance estimate is $df =$ _____ . Locate the F ratios associated with these two *df*'s in Table 3.

■ ■ ■ ■ ■ ■ ■ ■ ■ ■ ■ ■ ■ ■

12, 16

31. Table 3 indicates that, for these degrees of freedom, an F ratio of _____ is required for the .10 level of significance, and an F ratio of _____ is required at the .02 level.

■ ■ ■ ■ ■ ■ ■ ■ ■ ■ ■ ■ ■ ■

2.42, 3.55

32. PLATE 80. Your obtained F ratio of 2.27 _____ (does/does not) reach the .10 level of significance.

■ ■ ■ ■ ■ ■ ■ ■ ■ ■ ■ ■ ■ ■

does not

33. Since the F ratio of the variances in Plate 80 does not reach the .10 level of significance, you are justified in _____ (accepting/rejecting) the null hypothesis.

■ ■ ■ ■ ■ ■ ■ ■ ■ ■ ■ ■ ■ ■

accepting

EXERCISES

1. The following statistics were obtained in Exercise 2 of Set 18. Use Formula 20 to compute the F ratio for these data.

Between groups	$SS_b = 180$	$df_b = 2$
Within groups	$SS_w = 431$	$df_w = 27$
Total	$SS_t = 611$	$df_t = 29$

2. Evaluate the significance of the F test computed in Exercise 1, using Table 3. Use $P = .01$ as your acceptable level for significance. Do you reject the null hypothesis on the basis of the F test?

3. Below are the variance estimates of manual skills scores obtained from two samples of fourth-grade children. The hypothesis is that there is a difference in the variability of manual skills scores between the populations from which the samples were selected. Use Formula 31 to compute the F ratio for these data. Use Table 3 to evaluate the significance of the F ratio. Use $P = .02$ as the acceptable level for significance. Do you accept or reject the null hypothesis?

Group 1	$N = 25$	$s^2 = 215$
Group 2	$N = 41$	$s^2 = 72$

4. What does the symbol F represent?

SCATTER DIAGRAMS /
INTRODUCTION TO CORRELATION

The statistical techniques presented thus far describe frequency distributions or measure differences between sets of data in which raw scores were obtained for only one variable. Thus, each individual was measured on only one characteristic and statistical inferences were made regarding the distribution of this characteristic in the population.

Another very useful statistical technique available to the researcher allows him to measure the relationship between two sets of data obtained from the same sample, or data from two samples which have been "matched" on some basis. In such instances, the researcher obtains two frequency distributions of data and determines the degree to which the two distributions are related. The statistical term used to describe this relationship is *correlation*. Correlation is defined in this set and its degrees of freedom are discussed. The distinction between positive and negative correlations will be presented, along with a method for depicting the relationship of two frequency distributions graphically by plotting the raw scores in a *scatter diagram.*

SPECIFIC OBJECTIVES OF SET 20

At the conclusion of this set you will be able to:

(1) depict the relationship between two sets of scores by drawing a scatter diagram.

(2) describe what is meant by a positive correlation.

(3) describe what is meant by a negative correlation.

(4) describe what is meant by a perfect and a moderate correlation.

(5) describe what is meant by a zero correlation.

1. If you have obtained a set of spelling scores, it may be said that you have data on the *variable* of spelling scores. If you have obtained the mental ages of a group of children, you have data on the _____ of mental age.

■ ■ ■ ■ ■ ■ ■ ■ ■ ■ ■ ■ ■

variable

---

2. "Arithmetic test scores" is called a variable because, among a group of students, these scores will _____.

■ ■ ■ ■ ■ ■ ■ ■ ■ ■ ■ ■ ■

vary

---

3. Up to this point, you have been concerned with data for only one variable. If you obtain arithmetic test scores and achievement test scores for each member of a group, you have data on _____variables.

■ ■ ■ ■ ■ ■ ■ ■ ■ ■ ■ ■ ■

two

---

4. When you have data for two variables for a group of people, you may wish to know if there is a relationship between the two variables. For example, you may wish to determine if there is a _____ between arithmetic test scores and achievement test scores.

■ ■ ■ ■ ■ ■ ■ ■ ■ ■ ■ ■ ■

relationship

---

5. If there is a relationship between two variables, it may be said that the variables are *associated*. Thus, if the members of the group who get high arithmetic test scores also get high achievement test scores, this is an indication that these two variables are _____.

■ ■ ■ ■ ■ ■ ■ ■ ■ ■ ■ ■ ■

associated

Plate 81.

Child	Arithmetic Test Scores (X)	Achievement Test Scores (Y)
A	70	50
B	65	40
C	60	30
D	55	20
E	50	10

Scatter Diagram

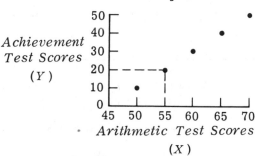

6. This plate presents the arithmetic test scores and the achievement test scores for five children. For child A, the arithmetic test score is 70 and the achievement test score is ____ .

■ ■ ■ ■ ■ ■ ■ ■ ■ ■ ■ ■ ■ ■

50

---

7. PLATE 81. Notice that the arithmetic test scores for the five children are listed in order of magnitude, with child A being the highest with 70, and child E being the lowest with ____ .

■ ■ ■ ■ ■ ■ ■ ■ ■ ■ ■ ■ ■ ■

50

8. PLATE 81. Notice also that the achievement test scores for the group are in the same order as the arithmetic test scores, with child A having 50, the highest score, and child E having ____, the lowest score.

■ ■ ■ ■ ■ ■ ■ ■ ■ ■ ■ ■ ■ ■

10

9. PLATE 81. On the variable of arithmetic test scores, there is a difference of 5 points between each of the scores. On the variable of achievement test scores, there is a difference of ____ points between each of the scores.

■ ■ ■ ■ ■ ■ ■ ■ ■ ■ ■ ■ ■ ■

10

10. PLATE 81. In this plate, there are two test scores for each child—an _____ test score and an _____ test score.

■ ■ ■ ■ ■ ■ ■ ■ ■ ■ ■ ■ ■ ■

arithmetic, achievement

11. PLATE 81. To get a picture of the relationship, or association, between these two variables, the data may be depicted graphically as is done at the bottom of the plate. Notice that the achievement test scores are listed on the vertical axis of the graph, and the arithmetic test scores are listed on the _____ axis.

■ ■ ■ ■ ■ ■ ■ ■ ■ ■ ■ ■ ■ ■

horizontal

12. PLATE 81. Notice that there are five dots in this graph. Each dot represents two scores for a child—an achievement test score and an _____ test score.

■ ■ ■ ■ ■ ■ ■ ■ ■ ■ ■ ■ ■ ■

arithmetic

13. PLATE 81. The location of each dot in the diagram designates two test scores for each child. Consider child D. His arithmetic test score is ____and his achievement test score is ____.

■ ■ ■ ■ ■ ■ ■ ■ ■ ■ ■ ■ ■ ■

55, 20

---

14. PLATE 81. It is general practice to designate all scores on the vertical axis as Y scores. Thus, in this plate, Y represents an _____ (achievement/arithmetic) score.

■ ■ ■ ■ ■ ■ ■ ■ ■ ■ ■ ■ ■ ■

achievement

---

15. PLATE 81. It is also general practice to designate all scores on the horizontal axis as X scores. Thus, in this plate, X represents an _____ (arithmetic/achievement) score.

■ ■ ■ ■ ■ ■ ■ ■ ■ ■ ■ ■ ■ ■

arithmetic

---

16. PLATE 81. Child D: X = 55, Y = ____.

■ ■ ■ ■ ■ ■ ■ ■ ■ ■ ■ ■ ■ ■

20

---

17. PLATE 81. Child D: X = 55, Y = 20.
The dotted lines in this diagram locate the position of child D. The achievement score (Y) is located directly across from score value 20 on the vertical axis. The arithmetic score (X) is located directly above score value ____ on the _____ axis.

■ ■ ■ ■ ■ ■ ■ ■ ■ ■ ■ ■ ■ ■

55, horizontal

---

18. PLATE 81. The one dot located at the intersection of the dotted lines represents both scores for child ___.

■ ■ ■ ■ ■ ■ ■ ■ ■ ■ ■ ■ ■ ■

D

---

19. PLATE 81. The dot representing child A is located directly across from score value _____ on the Y axis, and directly above score value _____ on the X axis.

■ ■ ■ ■ ■ ■ ■ ■ ■ ■ ■ ■ ■ ■

50, 70

---

20. PLATE 81. Notice that the five dots in this plate represent the arithmetic and achievement scores for each of the five children. Thus, one dot represents _____ scores for each child.

■ ■ ■ ■ ■ ■ ■ ■ ■ ■ ■ ■ ■ ■

two

---

21. PLATE 81. The diagram in this plate is called a *scatter diagram* because it indicates how the scores of the children _____ when they are plotted in this manner.

■ ■ ■ ■ ■ ■ ■ ■ ■ ■ ■ ■ ■ ■

scatter

---

22. PLATE 81. The scores on variable X are positively related, or associated, with the scores on variable Y in this plate, because, as X increases, Y also _____.

■ ■ ■ ■ ■ ■ ■ ■ ■ ■ ■ ■ ■ ■

increases

---

23. PLATE 81. When each increase in X scores is matched by a proportionate increase in Y scores, the two variables are said to be perfectly _____ .

■ ■ ■ ■ ■ ■ ■ ■ ■ ■ ■ ■ ■ ■

related

---

24. The statistical term for the amount of association between two variables is *correlation.* Thus, it can be said that there is a perfect _____ between X and Y scores in Plate 81.

■ ■ ■ ■ ■ ■ ■ ■ ■ ■ ■ ■ ■ ■

correlation

25. Two variables are said to be correlated when the scores on the two variables are associated. Thus, in Plate 81, there is a correlation between the children's arithmetic test scores and their_____test scores.

■ ■ ■ ■ ■ ■ ■ ■ ■ ■ ■ ■ ■ ■

achievement

26. PLATE 81. Notice that the dots in the scatter diagram fall in a straight line, from the lower left corner of the diagram to the upper _____ corner.

■ ■ ■ ■ ■ ■ ■ ■ ■ ■ ■ ■ ■ ■

right

27. PLATE 81. There is a positive correlation between the X and Y variables in this plate, because with each increase in X scores, there is a corresponding increase in____scores.

■ ■ ■ ■ ■ ■ ■ ■ ■ ■ ■ ■ ■ ■

Y

28. When the X scores of a scatter diagram are directly associated with the Y scores, it is called a _____ (positive/negative) correlation.

■ ■ ■ ■ ■ ■ ■ ■ ■ ■ ■ ■ ■ ■

positive

Plate 82.

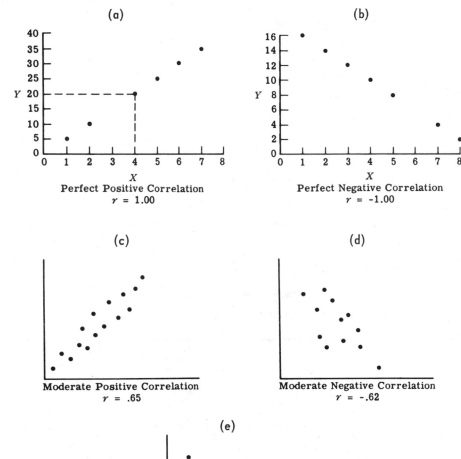

(a)

Perfect Positive Correlation
$r = 1.00$

(b)

Perfect Negative Correlation
$r = -1.00$

(c)

Moderate Positive Correlation
$r = .65$

(d)

Moderate Negative Correlation
$r = -.62$

(e)

Zero Correlation
$r = .00$

29. PLATE 82b. This plate presents another scatter diagram. The dots in this diagram run from the upper left corner to the _____ _____ corner.

■ ■ ■ ■ ■ ■ ■ ■ ■ ■ ■ ■ ■

lower right

30. PLATE 82b. In this diagram, the X scores are _____ (directly/ inversely) associated with the Y scores.

■ ■ ■ ■ ■ ■ ■ ■ ■ ■ ■ ■ ■ ■

inversely

31. PLATE 82b. The X scores are inversely related to the Y scores because as each X score increases, each Y score _____ .

■ ■ ■ ■ ■ ■ ■ ■ ■ ■ ■ ■ ■ ■

decreases

32. When the X and Y scores are directly related as in Plate 82a, the correlation is positive. When the X and Y scores are inversely related, as in Plate 82b, the correlation is _____ .

■ ■ ■ ■ ■ ■ ■ ■ ■ ■ ■ ■ ■ ■

negative.

33. When the dots lie in a straight line, the correlation is perfect. Thus, in Plate 82a, there is a perfect positive correlation. In Plate 82b, there is a perfect _____ correlation.

■ ■ ■ ■ ■ ■ ■ ■ ■ ■ ■ ■ ■ ■

negative

34. A perfect positive correlation exists whenever there is a(n) _____ (direct/inverse) perfect association between the X scores and the Y scores.

■ ■ ■ ■ ■ ■ ■ ■ ■ ■ ■ ■ ■ ■

direct

35. A perfect negative correlation exists whenever there is a(n) _____ (direct/inverse) perfect association between the X scores and the Y scores.

■ ■ ■ ■ ■ ■ ■ ■ ■ ■ ■ ■ ■ ■

inverse

36. Plate 82a is a scatter diagram of a perfect _____ correlation.

■ ■ ■ ■ ■ ■ ■ ■ ■ ■ ■ ■ ■ ■

positive

---

37. Plate 82b is a scatter diagram of a perfect _____ correlation.

■ ■ ■ ■ ■ ■ ■ ■ ■ ■ ■ ■ ■ ■

negative

---

38. Although correlations may be positive or negative, they generally are not perfect. Plate 82c presents a moderate _____ correlation.

■ ■ ■ ■ ■ ■ ■ ■ ■ ■ ■ ■ ■ ■

positive

---

39. PLATE 82c. (The values of X and Y are immaterial. Only the placement of the dots is important.) In this plate, every increase in X scores_____ (is/is not) matched by a corresponding increase in Y scores.

■ ■ ■ ■ ■ ■ ■ ■ ■ ■ ■ ■ ■ ■

is not

---

40. PLATE 82c. The general trend of the dots in this scatter diagram is from the lower left corner to the upper right corner; so the correlation, although not perfect, is a _____ (positive/negative) one.

■ ■ ■ ■ ■ ■ ■ ■ ■ ■ ■ ■ ■ ■

positive

---

41. PLATE 82c. This scatter diagram presents a moderate positive correlation because the data generally lie in a positive direction, although they_____ _____(do/do not) all lie on the diagonal of the diagram.

■ ■ ■ ■ ■ ■ ■ ■ ■ ■ ■ ■ ■ ■

do not

---

42. PLATE 82d. The general trend of the data in this scatter diagram is in a _____ (positive/negative) direction.

■ ■ ■ ■ ■ ■ ■ ■ ■ ■ ■ ■ ■ ■

negative

---

43. PLATE 82d. The dots lie in a generally negative direction but do not lie in a straight line. This _____ (is/is not) a perfect negative correlation.

■ ■ ■ ■ ■ ■ ■ ■ ■ ■ ■ ■ ■ ■

is not

---

44. PLATE 82e. This scatter diagram presents a zero correlation because there is no trend among the dots. In this diagram the scores on the X variable _____ (are/are not) associated with the scores on the Y variable.

■ ■ ■ ■ ■ ■ ■ ■ ■ ■ ■ ■ ■ ■

are not

---

45. PLATE 82e. A zero correlation indicates that there is _____ relationship between the scores on the X variable and the scores on the Y variable.

■ ■ ■ ■ ■ ■ ■ ■ ■ ■ ■ ■ ■

no

---

46. The scatter diagrams in Plate 82 indicate that correlations may range from perfect positive to perfect _____ correlations.

■ ■ ■ ■ ■ ■ ■ ■ ■ ■ ■ ■ ■ ■■

negative

1. Below are the final examination scores in algebra and English for eight students. Draw a scatter diagram of these scores. Let the algebra scores be the X variable.

Student	Algebra Scores	English Scores
A	81	93
B	84	97
C	86	98
D	82	94
E	85	96
F	82	95
G	83	94
H	84	95

2. What is meant by a perfect positive correlation between two variables? A moderate positive correlation?

3. What is meant by a perfect negative correlation between two variables? A moderate negative correlation?

4. Does the scatter diagram you prepared in Exercise 1 indicate a positive or a negative correlation? Is it a perfect correlation?

5. What is meant by a zero correlation?

PEARSON PRODUCT-MOMENT CORRELATION

To express the relationship between two variables statistically, we must have some numerical index of the degree of correlation. This index is termed *correlation coefficient*, and its magnitude indicates the degree to which two frequency distributions of data are related.

There are many different types of correlation analysis available to the researcher. Which one is appropriate for his use depends upon the nature of the data he is analyzing. This set presents a method for computing one type of correlation coefficient—the Pearson product-moment correlation coefficient. This is one of the more common correlation techniques. In order to use it properly, however, the researcher must assume that the variables are linearly related and the scores on each variable are normally distributed. If these assumptions cannot be made, this type of correlation analysis is inappropriate and other techniques must be used. Set 23 of this book describes a correlation technique which can be used for data not meeting the above assumptions.

This set presents the method for evaluating the significance of the Pearson product-moment correlation coefficient by using Table 4, and the basis for determining whether to accept or reject the null hypothesis. We will also discuss one-and two-tailed tests of significance for correlation coefficients.

SPECIFIC OBJECTIVES OF SET 21

At the conclusion of this set you will be able to:

(1) state the coefficient which denotes a perfect positive correlation.

(2) state the coefficient which denotes a perfect negative correlation.

(3) use Formula 32 to calculate a Pearson product-moment correlation coefficient when given data on two variables for the same sample.

(4) state the null hypothesis regarding correlated variables.

(5) use Table 4 to determine the level of significance of a Pearson product-moment correlation coefficient.

(6) make a research hypothesis which requires you to use a one-tailed test of significance.

(7) use Table 4 for determining significance levels for one- and for two-tailed tests of significance.

1. In statistics, the degree of correlation between two variables is indicated by a *correlation coefficient*. The most commonly used correlation coefficient is called the Pearson product-moment correlation _____.

■ ■ ■ ■ ■ ■ ■ ■ ■ ■ ■ ■ ■ ■

coefficient

---

2. The symbol for the Pearson product-moment correlation coefficient is r. This coefficient indicates the degree of _____ between two variables.

■ ■ ■ ■ ■ ■ ■ ■ ■ ■ ■ ■ ■ ■

correlation

---

3. PLATE 82a, page 262. As indicated in this plate, the Pearson product-moment coefficient that indicates a perfect positive correlation is $r =$ _____.

■ ■ ■ ■ ■ ■ ■ ■ ■ ■ ■ ■ ■ ■

1.00

---

4. PLATE 82b. A perfect negative correlation has the value of $r =$ _____.

■ ■ ■ ■ ■ ■ ■ ■ ■ ■ ■ ■ ■ ■

−1.00

---

5. When no correlation exists between two variables, $r =$ _____.

■ ■ ■ ■ ■ ■ ■ ■ ■ ■ ■ ■ ■ ■

.00

---

6. PLATE 82. As indicated by the scatter diagrams, the possible range of correlation coefficients is from −1.00 to _____.

■ ■ ■ ■ ■ ■ ■ ■ ■ ■ ■ ■ ■ ■

1.00

7. PLATE 82c. The correlation coefficient for this scatter diagram is $r =$ _____. This indicates a moderate positive correlation.

■ ■ ■ ■ ■ ■ ■ ■ ■ ■ ■ ■ ■ ■

.65

8. PLATE 82d. The correlation coefficient for this scatter diagram is $r = -.62$. This indicates a moderate _____ correlation.

■ ■ ■ ■ ■ ■ ■ ■ ■ ■ ■ ■ ■ ■

negative

9. All correlation coefficients that lie between 1.00 and .00 indicate a positive relationship. Correlation coefficients that lie between .00 and -1.00 indicate a _____ relationship.

■ ■ ■ ■ ■ ■ ■ ■ ■ ■ ■ ■ ■ ■

negative

10. A correlation coefficient of $r = .70$ indicates a _____ (smaller/ greater) degree of association than $r = .50$.

■ ■ ■ ■ ■ ■ ■ ■ ■ ■ ■ ■ ■ ■

greater

11. Correlation coefficients of $r = .80$ and $r = -.80$ indicate the *same degree* of association: $r = .80$ indicates a positive association, whereas $r = -.80$ indicates a _____ association.

■ ■ ■ ■ ■ ■ ■ ■ ■ ■ ■ ■ ■ ■

negative

12. Correlation coefficients of $r = -.20$ and $r = .20$ indicate _____ (differing/the same) degree of association.

■ ■ ■ ■ ■ ■ ■ ■ ■ ■ ■ ■ ■ ■

the same

13. Correlation coefficients range from _____ to _____ .

−1.00, 1.00

Formula 32. Calculation of the Pearson Product-Moment Correlation Coefficient

$$r = \frac{N\Sigma XY - (\Sigma X)(\Sigma Y)}{\sqrt{[N\Sigma X^2 - (\Sigma X)^2][N\Sigma Y^2 - (\Sigma Y)^2]}}$$

N = number of pairs of scores

Degrees of Freedom: $N - 2$

---

14. This formula is to be used for the calculation of the Pearson _____ _____ correlation coefficient.

product-moment

---

15. FORMULA 32. The formula is used to determine the correlation coefficient between two variables. The symbol for a raw score on one variable is X. The symbol for a raw score on the other variable is ___ (symbol).

Y

---

16. FORMULA 32. The expression ΣXY indicates that you multiply the X score times the Y score for each individual, and then _____ all the products.

sum

---

17. FORMULA 32. The expression $N\Sigma XY$ indicates that you multiply the sum of the products by ___ (symbol).

N

---

18. FORMULA 32. The expression $(\Sigma X)(\Sigma Y)$ indicates that you sum the X scores, and sum the Y scores, and then _____ these two sums.

■ ■ ■ ■ ■ ■ ■ ■ ■ ■ ■ ■ ■ ■

multiply

---

19. FORMULA 32. The expression $N\Sigma X^2$ indicates that you _____the squares of the X scores and multiply by___ (symbol).

■ ■ ■ ■ ■ ■ ■ ■ ■ ■ ■ ■ ■ ■

sum, N

---

20. FORMULA 32. The expression $(\Sigma X)^2$ indicates that you sum the X scores and then _____ the sum.

■ ■ ■ ■ ■ ■ ■ ■ ■ ■ ■ ■ ■ ■

square

---

21. The expression ΣX^2 and $(\Sigma X)^2$ differ in that ΣX^2 indicates that you square the scores first and then _____ them, while $(\Sigma X)^2$ indicates that you sum the scores first and _____ the sum.

■ ■ ■ ■ ■ ■ ■ ■ ■ ■ ■ ■ ■ ■

sum, square

---

22. PLATE 82a. This is a scatter diagram of a perfect positive correlation. The scores plotted here are presented in Plate 83. To compute the value of r using Formula 32, you must determine six values: N, ΣX, ΣY, ΣX^2, _____ , and _____.

■ ■ ■ ■ ■ ■ ■ ■ ■ ■ ■ ■ ■ ■

ΣY^2, ΣXY

Plate 83.

	(Col. 1)	(Col. 2)	(Col. 3)	(Col. 4)	(Col. 5)
Child	X	Y	X^2	Y^2	XY
A	1	5	1	25	5
B	2	10	4	100	20
C	4	20	16	400	80
D	5	25	25	625	125
E	6	30	36	900	180
F	7	35	49	1225	245
$N = 6$	$\Sigma X = 25$	$\Sigma Y = 125$	$\Sigma X^2 = 131$	$\Sigma Y^2 = 3275$	$\Sigma XY = 655$

(Formula 32) $\quad r = \dfrac{N\Sigma XY - (\Sigma X)(\Sigma Y)}{\sqrt{[N\Sigma X^2 - (\Sigma X)^2][N\Sigma Y^2 - (\Sigma Y)^2]}}$

$$= \frac{6(655) - (25)(125)}{\sqrt{[6(131) - (25)^2][6(3275) - (125)^2]}}$$

$$= \frac{805}{\sqrt{(161)(4025)}} = \frac{805}{\sqrt{648,025}} = \frac{805}{805} = 1.000$$

23. In this example, N = ___. Small numbers of scores are used for illustration only. Of course, the same procedures apply for any number of scores.

■ ■ ■ ■ ■ ■ ■ ■ ■ ■ ■ ■ ■ ■

6

---

24. PLATE 83. Each X score is presented in column 1. Each Y score is presented in column 2. The square of each X score is presented in column 3. The square of each Y score is presented in column ___ .

■ ■ ■ ■ ■ ■ ■ ■ ■ ■ ■ ■ ■ ■

4

---

25. PLATE 83. The product of each X score with its Y score is represented by the symbol XY and is presented in column ___ .

■ ■ ■ ■ ■ ■ ■ ■ ■ ■ ■ ■ ■ ■

5

---

26. PLATE 83. The sum of each of the columns in the plate gives you the values needed for substitution in Formula 32. In this example: $N = 6$, $\Sigma X = 25$, $\Sigma Y = 125$, $\Sigma X^2 = $ _____ , $\Sigma Y^2 = $ _____ , and $\Sigma XY = $ _____ .

■ ■ ■ ■ ■ ■ ■ ■ ■ ■ ■ ■ ■

131, 3275, 655

27. PLATE 83. The numerator of Formula 32 is $N\Sigma XY - (\Sigma X)(\Sigma Y)$, which indicates that you multiply N times ΣXY and subtract the product of _____(symbol) and _____(symbol).

■ ■ ■ ■ ■ ■ ■ ■ ■ ■ ■ ■ ■

$(\Sigma X), (\Sigma Y)$

28. PLATE 83. The denominator of Formula 32 indicates that you do the necessary computation within the brackets and then multiply the two figures. Then you extract the _____ _____ of the product.

■ ■ ■ ■ ■ ■ ■ ■ ■ ■ ■ ■ ■

square root

29. PLATE 83. Check the substitution of the numerical values for the symbols in Formula 32. Also, check the arithmetic. In this example, $r = 1.00$, which indicates a perfect _____ correlation.

■ ■ ■ ■ ■ ■ ■ ■ ■ ■ ■ ■ ■

positive

Plate 84.

Person	X	Y
A	4	9
B	1	4
C	3	1
D	6	7
E	5	3
F	4	2

30. This plate presents data for six persons, each having an X score and a Y score. Prepare a scatter diagram for these data.

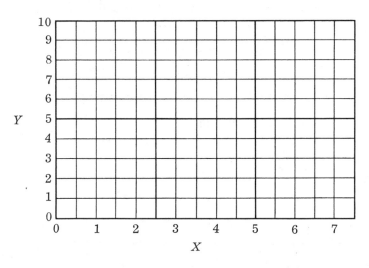

■ ■ ■ ■ ■ ■ ■ ■ ■ ■ ■ ■ ■

Plate 85.

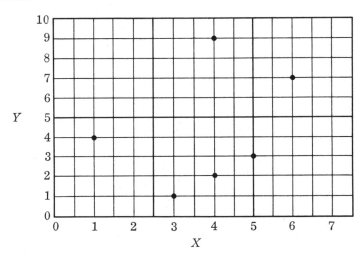

31. PLATE 85. This scatter diagram of the data in Plate 84, although far from representing a perfect correlation, indicates a slight _____ (positive/negative) relationship.

■ ■ ■ ■ ■ ■ ■ ■ ■ ■ ■ ■ ■

positive

32. PLATE 85. The value of r ranges from -1.00 to 1.00. If there is a slight positive correlation between X and Y in this plate, the value of r must lie somewhere between the values of _____ and _____.

■ ■ ■ ■ ■ ■ ■ ■ ■ ■ ■ ■ ■ ■

.00, 1.00

---

33. PLATE 84. In order to compute the value of r using the column headings of X, Y, X^2, Y^2, and XY (as in Plate 83), do the necessary arithmetic to determine the following: $N =$ ___ , $\Sigma X =$ ____ , $\Sigma Y =$ ____ , $\Sigma X^2 =$ _____ , $\Sigma Y^2 =$ _____ , and $\Sigma XY =$ _____ .

■ ■ ■ ■ ■ ■ ■ ■ ■ ■ ■ ■ ■ ■

6, 23, 26, 103, 160, 108

---

34. PLATE 84. $N = 6$, $\Sigma X = 23$, $\Sigma Y = 26$, $\Sigma X^2 = 103$, $\Sigma Y^2 = 160$, $\Sigma XY = 108$. Substitute these values for the symbols in Formula 32 and do the necessary calculations to determine the value of r. $r =$ _____ . (to two decimal places)

■ ■ ■ ■ ■ ■ ■ ■ ■ ■ ■ ■ ■ ■

$r = .31$

(see below to check your computation)

Plate 86.

Child	X	Y	X^2	Y^2	XY
A	4	9	16	81	36
B	1	4	1	16	4
C	3	1	9	1	3
D	6	7	36	49	42
E	5	3	25	9	15
F	4	2	16	4	8
$N = 6$	$\Sigma X = 23$	$\Sigma Y = 26$	$\Sigma X^2 = 103$	$\Sigma Y^2 = 160$	$\Sigma XY = 108$

$$r = \frac{N\Sigma XY - (\Sigma X)(\Sigma Y)}{\sqrt{[N\Sigma X^2 - (\Sigma X)^2][N\Sigma Y^2 - (\Sigma Y)^2]}}$$

$$= \frac{6(108) - (23)(26)}{\sqrt{[6(103) - (23)^2][6(160) - (26)^2]}}$$

$$= \frac{50}{\sqrt{(89)(284)}} = \frac{50}{\sqrt{25276}} = \frac{50}{158.98} = .31$$

---

35. PLATE 86. $r = .31$. This indicates that there is a slight positive correlation between the X and Y variables for the six people. If you had obtained a value of $r = .60$, this would have indicated a _____(greater/lesser) correlation between the two variables.

■ ■ ■ ■ ■ ■ ■ ■ ■ ■ ■ ■ ■ ■

greater

---

36. As indicated in the two examples that have been presented, when, in the numerator of Formula 32, $N\Sigma XY$ is greater than $(\Sigma X)(\Sigma Y)$, the correlation is positive. When $N\Sigma XY$ is less than $(\Sigma X)(\Sigma Y)$, the correlation must be

_____ .

■ ■ ■ ■ ■ ■ ■ ■ ■ ■ ■ ■ ■ ■

negative

---

37. FORMULA 32. When the numerator term, $N\Sigma XY - (\Sigma X)(\Sigma Y) = 0$, then r must equal ____ .

■ ■ ■ ■ ■ ■ ■ ■ ■ ■ ■ ■ ■

.00

---

38. In testing hypotheses regarding correlations, the null hypothesis states, "There is ____ correlation between variable X and variable Y."

■ ■ ■ ■ ■ ■ ■ ■ ■ ■ ■ ■ ■ ■

no

---

39. PLATE 86. The correlation between variables X and Y is r = .31. You need to determine if this value is sufficiently large to permit you to reject the _____ hypothesis.

■ ■ ■ ■ ■ ■ ■ ■ ■ ■ ■ ■ ■

null

---

40. PLATE 86. To evaluate the significance of r, use Table 4. This table presents the values of r that are required for P = .10, P = .05, P = _____ , and P = _____ for differing degrees of freedom (df).

■ ■ ■ ■ ■ ■ ■ ■ ■ ■ ■ ■ ■

.02, .01

---

41. To determine the value of r that is significant for the differing P values, you must first determine the df for the data. Formula 32 indicates that the df to be used in evaluating r is df = _____ .

■ ■ ■ ■ ■ ■ ■ ■ ■ ■ ■ ■ ■

$N-2$

---

42. FORMULA 32. To evaluate the significance of r, the df = $N-2$. In this case, N is the number of individuals involved in the correlation. Thus, N is the number of pairs of scores. In Plate 86, df = _____.

■ ■ ■ ■ ■ ■ ■ ■ ■ ■ ■ ■ ■

4

---

43. PLATE 86. r = .31, df = 4. From Table 4, for df = 4, the value of r necessary for significance at the .05 level is____ .

■ ■ ■ ■ ■ ■ ■ ■ ■ ■ ■ ■ ■

.81

44. PLATE 86. $r = .31$, $df = 4$. From Table 4, significance at the .05 level would require an r of .81. Therefore, because the obtained r is only .31, you _____ (accept/reject) the null hypothesis that there is no correlation between the X and Y variables.

■ ■ ■ ■ ■ ■ ■ ■ ■ ■ ■ ■ ■ ■

accept

---

45. PLATE 86. $r = .31$, $df = 4$. From Table 4; for $P = .05$, $r = .81$. You would accept the null hypothesis that there is no correlation between the two variables. This means that you conclude that the obtained r of .31 might well be due to _____ error.

■ ■ ■ ■ ■ ■ ■ ■ ■ ■ ■ ■ ■ ■

sampling

---

46. PLATE 86. $r = .31$, $df = 4$. Table 4. If the df for these data had been 50, then the required r for $P = .05$ would be _____ .

■ ■ ■ ■ ■ ■ ■ ■ ■ ■ ■ ■ ■ ■

.273

---

47. Consider: $r = .31$, $df = 50$. From Table 4, the r required for significance at the .05 level is .273. Therefore, an obtained r of .31 would be considered _____ (significant/non-significant).

■ ■ ■ ■ ■ ■ ■ ■ ■ ■ ■ ■ ■ ■

significant

---

48. Consider: $r = .31$, $df = 50$. An obtained r of .31 is significant beyond the .05 level because this r is _____ (larger/smaller) than the required value of r at the .05 level as presented in Table 4.

■ ■ ■ ■ ■ ■ ■ ■ ■ ■ ■ ■ ■ ■

larger

49. Consider: $r = .31$, $df = 50$. Because r is beyond the .05 level of significance, it may be said that, if you reject the null hypothesis, the probability of a Type I error is _____ (less/greater) than $P = .05$.

■ ■ ■ ■ ■ ■ ■ ■ ■ ■ ■ ■ ■ ■

less

---

50. If, for a particular df, the obtained r exceeds the tabled value of r at $P = .05$, the correlation is said to be significant beyond the .05 level. If the obtained r exceeds the tabled value of r at $P = .01$, the correlation is said to be significant beyond the _____ level.

■ ■ ■ ■ ■ ■ ■ ■ ■ ■ ■ ■ ■ ■

.01

---

51. Table 4 is for two-tailed significance tests. Therefore, the values in this table apply to both positive and _____ correlations.

■ ■ ■ ■ ■ ■ ■ ■ ■ ■ ■ ■ ■ ■

negative

---

52. If you make the research hypothesis that your correlation will be positive, you have made a _____ (one/two) tailed hypothesis.

■ ■ ■ ■ ■ ■ ■ ■ ■ ■ ■ ■ ■ ■

one

---

53. If, in your research hypothesis, you do not state whether the correlation is positive or negative, you should use a _____ (one/two) tailed test of significance.

■ ■ ■ ■ ■ ■ ■ ■ ■ ■ ■ ■ ■ ■

two

54. If you make a one-tailed hypothesis, you may make a one-tailed test of significance. This is done in exactly the same manner as with the t test.

TABLE 4. For a one-tailed test, the values of r that will be significant at the .05 level are listed in the column headed P = _____ .

■ ■ ■ ■ ■ ■ ■ ■ ■ ■ ■ ■ ■

.10

---

55. TABLE 4. For a one-tailed test, the r's at the P = .01 level are listed in the column headed P = _____ .

■ ■ ■ ■ ■ ■ ■ ■ ■ ■ ■ ■ ■

.02

EXERCISES

1. What magnitude must a Pearson product-moment correlation coefficient have to denote a perfect positive correlation?

2. For the following set of data, the researcher has made this hypothesis "There is a relationship between the arithmetic scores and the English scores." Use Formula 32 to compute the Pearson product-moment correlation coefficient for these data.

Student	Arithmetic Scores	English Scores
A	20	18
B	18	22
C	17	15
D	16	17
E	14	8
F	14	20
G	12	9
H	9	7

3. Does the hypothesis stated in Exercise 2 require a one-tailed or a two-tailed test of significance? Using Table 4, determine the magnitude of the r required for significance at the .05 level. At the .01 level.

4. State the null hypothesis being tested in Exercise 2. Using $P = .05$ as your acceptable level for significance, would you accept or reject the null hypothesis?

5. For the following set of data, the researcher has made this hypothesis: "There is a positive relationship between the spelling scores and the English scores." Compute the Pearson product-moment correlation coefficient for these data, using Formula 32.

Student	Spelling Scores	English Scores	Student	Spelling Scores	English Scores
A	32	20	G	25	10
B	29	17	H	25	8
C	28	17	I	21	9
D	27	18	J	20	6
E	27	17	K	15	8
F	27	12			

6. Does the hypothesis stated in Exercise 5 require a one-tailed or a two-tailed test of significance? Using Table 4, determine the magnitude of the r required for significance at the .05 level. At the .01 level.

7. State the null hypothesis being tested in Exercise 5. If you designate $P = .01$ as your level of significance, would you accept or reject the null hypothesis?

8. Identify the symbol r.

REGRESSION

If arithmetic scores and intelligence test scores are correlated, it is possible to estimate the value of a person's arithmetic score on the basis of his intelligence test score. Likewise, if a person's arithmetic score is known, it is possible to estimate his intelligence test score.

This set demonstrates the use of the *regression coefficient* in making such predictions, and introduces and explains the new terms *regression line*, *best fit line*, and *regression equation*. A method is given for plotting regression lines for both variables, using the regression equations, and there is a discussion of the relationship between regression lines in cases where the two variables are not perfectly correlated.

SPECIFIC OBJECTIVES OF SET 22

At the conclusion of this set you will be able to:

(1)   state what is meant by *regression* or *best fit line*.

(2)   use Formula 33 to calculate a regression equation for the regression of Y on X.

(3)   use Formula 36 to calculate a regression equation for the regression of X on Y.

(4) plot the regression lines in a scatter diagram.

(5) predict, using the regression equation, a score on one variable that is associated with a given score on the other variable.

(6) identify the symbols $\tilde{Y}$, b_{yx}, $\tilde{X}$, b_{xy}.

1. If two variables are correlated, you can estimate the value of the score on one variable for an individual if you know his score on the other variable. Thus, if you know one child's achievement test score, it is possible for you to _____ his arithmetic test score.

■ ■ ■ ■ ■ ■ ■ ■ ■ ■ ■ ■ ■ ■

estimate

---

2. When you have two correlated variables, it is possible to estimate the Y score of an individual on the basis of his X score. If height and age are correlated, you may estimate an individual's height if you know his _____.

■ ■ ■ ■ ■ ■ ■ ■ ■ ■ ■ ■ ■ ■

age

Plate 87. Scatter Diagram with the Regression Line of Y on X

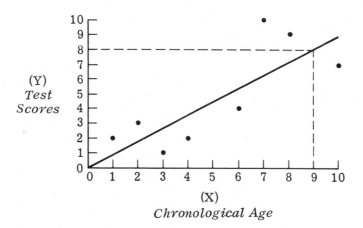

3. This plate presents the scatter diagram of test scores for eight children at various ages. For example: the child who is one year old has a test score of 2; the child who is six years old has a test score of____.

■ ■ ■ ■ ■ ■ ■ ■ ■ ■ ■ ■ ■ ■

4

4. PLATE 87. The trend of the data in this plate indicates that there is a _____ (positive/negative) correlation between these two variables.

■ ■ ■ ■ ■ ■ ■ ■ ■ ■ ■ ■ ■ ■

positive

5. PLATE 87. The line that is shown in this plate is called a *regression line*. This particular line shows the regression of variable Y (test scores) on variable X (_____ _____).

■ ■ ■ ■ ■ ■ ■ ■ ■ ■ ■ ■ ■ ■

chronological age

6. PLATE 87. Although the dots representing the test scores at the various ages do not lie exactly on this line, it is the one straight line that best fits the data. Therefore, this line is called the *regression line* of Y on X, or the _____ __ *line*.

■ ■ ■ ■ ■ ■ ■ ■ ■ ■ ■ ■ ■ ■

best fit

7. PLATE 87. This best fit, or regression, line can be used when you wish to estimate the value of a test score (Y) for a child at a particular chronological age (___) (symbol).

■ ■ ■ ■ ■ ■ ■ ■ ■ ■ ■ ■ ■ ■

X

8. PLATE 87. Although you do not have a nine-year old child in your sample, it is possible to estimate the score value for that age child by using the best fit or _____ line.

■ ■ ■ ■ ■ ■ ■ ■ ■ ■ ■ ■ ■ ■

regression

9. PLATE 87. Thus, using the regression line, the best estimate of the test score for a nine-year-old child is ___ . (See dotted lines.)

■ ■ ■ ■ ■ ■ ■ ■ ■ ■ ■ ■ ■ ■

8

---

10. PLATE 87. The six-year-old child in the sample had a test score of ___. However, based on the regression line, the best estimate of the test score for all six-year-old children is _____ .

■ ■ ■ ■ ■ ■ ■ ■ ■ ■ ■ ■ ■ ■

4, 5.5

---

11. Although the regression line is drawn using sample data, it enables you to predict the most likely value of test scores for the _____ (sample/population).

■ ■ ■ ■ ■ ■ ■ ■ ■ ■ ■ ■ ■ ■

population

---

12. Thus, using the regression line, it is possible to determine the best estimate of a value on variable Y for any particular value on variable ___ (symbol).

■ ■ ■ ■ ■ ■ ■ ■ ■ ■ ■ ■ ■ ■

X

Formulas 33 through 35. Formulas for Calculation of the Regression of Y on X

Formula 33 $\qquad \widetilde{Y} = a + b_{yx}X \qquad$ (regression equation)

Formula 34 $\qquad b_{yx} = \dfrac{\Sigma XY - \dfrac{(\Sigma X)(\Sigma Y)}{N}}{\Sigma X^2 - \dfrac{(\Sigma X)^2}{N}} \qquad$ (regression coefficient)

Formula 35 $\qquad a = \overline{Y} - b_{yx}\overline{X}$

in which $\widetilde{Y}$ is the predicted value of Y

13. The position of the regression line in a scatter diagram is determined algebraically by use of Formula 33. This formula is called

_____ _____ .

■ ■ ■ ■ ■ ■ ■ ■ ■ ■ ■ ■ ■ ■

regression equation

14. FORMULA 33. The symbol $\tilde{Y}$ is used to indicate the _____ value of Y that corresponds with any particular score value of X.

■ ■ ■ ■ ■ ■ ■ ■ ■ ■ ■ ■ ■ ■

predicted

15. FORMULA 34. The symbol b_{yx} is called the regression coefficient. The subscript, yx, in this symbol indicates that this is the regression coefficient for variable Y on variable ____ (symbol).

■ ■ ■ ■ ■ ■ ■ ■ ■ ■ ■ ■ ■ ■

X

16. FORMULA 34. In this formula for b_{yx} the numerator is divided by the sum of _____ of the X variable. (Recall Formula 11b).

■ ■ ■ ■ ■ ■ ■ ■ ■ ■ ■ ■ ■ ■

squares

17. FORMULA 35. The symbol a is used to represent a *constant* to be used in Formula 33. It is obtained by multiplying b_{yx} times $\bar{X}$ and subtracting the product from ____ (symbol).

■ ■ ■ ■ ■ ■ ■ ■ ■ ■ ■ ■ ■ ■

$\bar{Y}$

Plate 88. Computation of the Regression Equation of Y on X

Individual	Chronological Ages X	Test Scores Y	X^2	Y^2	XY
A	10	7	100	49	70
B	8	9	64	81	72
C	7	10	49	100	70
D	6	4	36	16	24
E	4	2	16	4	8
F	3	1	9	1	3
G	2	3	4	9	6
H	1	2	1	4	2
$N = 8$	$\Sigma X = 41$	$\Sigma Y = 38$	$\Sigma X^2 = 279$	$\Sigma Y^2 = 264$	$\Sigma XY = 255$

Formula 34.

$$b_{yx} = \frac{\Sigma XY - \dfrac{(\Sigma X)(\Sigma Y)}{N}}{\Sigma X^2 - \dfrac{(\Sigma X)^2}{N}} = \frac{255 - \dfrac{(41)(38)}{8}}{279 - \dfrac{(41)^2}{8}} = \frac{60.25}{68.88} = .87$$

$$\bar{X} = \frac{41}{8} = 5.12 \qquad \bar{Y} = \frac{38}{8} = 4.75$$

Formula 35.

$$a = \bar{Y} - b_{yx}\bar{X} = 4.75 - (.87)(5.12) = .30$$

18. This plate presents the data plotted in the scatter diagram in Plate 87. For these data, $N =$ ___ .

8

290

19. **PLATE 88.** To determine the values to be used in Formula 33 for the regression equation, first you must determine the value of b_{yx}, which is called the _____ _____ .

■ ■ ■ ■ ■ ■ ■ ■ ■ ■ ■ ■ ■ ■ ■

regression coefficient

---

20. **PLATE 88.** To solve for b_{yx}, first you must substitute the numerical values for the symbols in Formula 34. Check the substitution in the formula with the values obtained from the columns at the top of the plate. In this plate, b_{yx} = _____ .

■ ■ ■ ■ ■ ■ ■ ■ ■ ■ ■ ■ ■ ■ ■

.87

---

21. **PLATE 88.** b_{yx} = .87. To determine the value of a, use Formula 35. In order to determine a, you must first compute the mean of each variable. Check these calculations in Plate 86. $\bar{X}$ = _____ , $\bar{Y}$ = _____ .

■ ■ ■ ■ ■ ■ ■ ■ ■ ■ ■ ■ ■ ■ ■

5.12, 4.75

---

22. **PLATE 88.** b_{yx} = .87, $\bar{X}$ = 5.12, $\bar{Y}$ = 4.75. To solve for a, substitute the numerical values for the symbols in Formula 35. Check these substitutions in Plate 88. a = _____ .

■ ■ ■ ■ ■ ■ ■ ■ ■ ■ ■ ■ ■ ■ ■

.30

---

23. **PLATE 88.** a = .30, b_{yx} = .87. With this information, you may now determine the regression equation for this set of data. Substitute in Formula 33 the numerical values for the symbols a and b_{yx}. $\tilde{Y}$ = _____ + _____ X.

■ ■ ■ ■ ■ ■ ■ ■ ■ ■ ■ ■ ■ ■ ■

.30, .87

---

24. PLATE 88. The regression equation for this set of data is $\widetilde{Y}$ = .30 + .87X. In order to make the best prediction of a score on the Y variable ($\widetilde{Y}$) corresponding to a score on the X variable, you must substitute the value of X in Formula 33 and solve for ____ (symbol).

■ ■ ■ ■ ■ ■ ■ ■ ■ ■ ■ ■ ■ ■

$\widetilde{Y}$

25. PLATE 88. $\widetilde{Y}$ = .30 + .87X. Suppose you wish to predict the test score value ($\widetilde{Y}$) for a five-year-old child (i.e., X = 5). Substitute 5 for X in Formula 33 and solve for $\widetilde{Y}$. $\widetilde{Y}$ = _____.

■ ■ ■ ■ ■ ■ ■ ■ ■ ■ ■ ■ ■ ■

$\widetilde{Y}$ = .30 + (.87)5 = 4.65

26. $\widetilde{Y}$ = .30 + .87X. PLATE 88. For X = 5, $\widetilde{Y}$ = 4.65. Thus, using the data in the plate, the best estimate or prediction of the test score for a five-year-old child is 4.65. Using Formula 33, determine the best estimate or prediction of the test score for a ten-year-old child. $\widetilde{Y}$ = ____. (to two decimal places)

■ ■ ■ ■ ■ ■ ■ ■ ■ ■ ■ ■ ■ ■

9.00   [.30 + (.87)10]

27. The position of the regression line is determined by calculating two values of Y, plotting them on the scatter diagram, and drawing a straight line through them. Notice, in Plate 87, that the regression line is determined by the values you have just determined: for X = 5, $\widetilde{Y}$ = 4.65; for X = 10, $\widetilde{Y}$ = 9.00. This line represents the regression of Y on ___(symbol).

■ ■ ■ ■ ■ ■ ■ ■ ■ ■ ■ ■ ■ ■

X

28. FORMULA 33. This formula is for the regression of Y on X. It is also possible to determine the _____ of X on Y.

■ ■ ■ ■ ■ ■ ■ ■ ■ ■ ■ ■ ■ ■

regression

Formulas 36 through 38. Formulas for Calculation of the Regression of X on Y

Formula 36 $\qquad \tilde{X} = a + b_{xy}Y$ $\qquad$ (regression equation)

Formula 37 $\qquad b_{xy} = \dfrac{\Sigma XY - \dfrac{(\Sigma X)(\Sigma Y)}{N}}{\Sigma Y^2 - \dfrac{(\Sigma Y)^2}{N}}$ $\qquad$ (regression coefficient)

Formula 38 $\qquad a = \bar{X} - b_{xy}\bar{Y}$

in which $\tilde{X}$ is the predicted value of X

29. Formula 36 presents the formula for determining the regression of____ (symbol) on ____(symbol).

■ ■ ■ ■ ■ ■ ■ ■ ■ ■ ■ ■ ■ ■

X, Y

---

30. FORMULA 36. In this formula the symbol $\tilde{X}$ is used to indicate the _____ value of X that corresponds with any particular score value of Y.

■ ■ ■ ■ ■ ■ ■ ■ ■ ■ ■ ■ ■ ■

predicted

---

31. FORMULA 34. In b_{yx} the subscript yx is used to indicate that this is the regression coefficient of Y on X. FORMULA 37. In b_{xy} the subscript xy is used to indicate that this is the regression coefficient of ____ (symbol) on ____ (symbol).

■ ■ ■ ■ ■ ■ ■ ■ ■ ■ ■ ■ ■ ■

X, Y

---

32. FORMULA 37. The regression coefficient b_{xy} for the regression of X on Y has as its divisor the sum of _____ of the Y variable.

■ ■ ■ ■ ■ ■ ■ ■ ■ ■ ■ ■ ■ ■

squares

---

33. In order to determine the regression coefficient of X on Y for the data in Plate 88, substitute the numerical values for the symbols in Formula 37 and solve for b_{xy}. b_{xy} = _____ (to two decimal places).

■ ■ ■ ■ ■ ■ ■ ■ ■ ■ ■ ■ ■ ■

$$\frac{60.25}{264 - \frac{(38)^2}{8}} = .72$$

---

34. PLATE 88. b_{xy} = .72. You have already determined the means of the two variables. From the plate, these are: $\bar{X}$ = _____ , $\bar{Y}$ = _____ .

■ ■ ■ ■ ■ ■ ■ ■ ■ ■ ■ ■ ■ ■

5.12, 4.75

---

35. PLATE 88. b_{xy} = .72, $\bar{X}$ = 5.12, $\bar{Y}$ = 4.75. FORMULA 38. Substitute the numerical values for the symbols in this formula and solve for a. a = _____ (to two decimal places).

■ ■ ■ ■ ■ ■ ■ ■ ■ ■ ■ ■ ■ ■

$5.12 - (.72)4.75 = 1.70$

---

36. PLATE 88. a = 1.70, b_{xy} = .72. FORMULA 36. Substitute the numerical values for the symbols in this formula and write the regression equation for X on Y for these data.

■ ■ ■ ■ ■ ■ ■ ■ ■ ■ ■ ■ ■ ■

$\tilde{X}$ = 1.70 + .72Y

37. The regression equation for X on Y for the data in Plate 88 is $\widetilde{X}$ = 1.70 + .72Y. Using this equation you can predict the value of ___(symbol) for any particular value of ___(symbol).

■ ■ ■ ■ ■ ■ ■ ■ ■ ■ ■ ■ ■ ■ ■

X, Y

38. PLATE 88. $\widetilde{X}$ = 1.70 + .72Y. For the score value Y = 3 determine the predicted value of X. For Y = 3, $\widetilde{X}$ = _____ (to two decimal places).

■ ■ ■ ■ ■ ■ ■ ■ ■ ■ ■ ■ ■ ■ ■

3.86   [1.70 + .72(3)]

39. PLATE 88. $\widetilde{X}$ = 1.70 + .72Y. For Y = 3, $\widetilde{X}$ = 3.86. This means that if an individual receives a score of 3 on variable Y, you predict, from the regression equation, that on variable X he will receive a score of _____ .

■ ■ ■ ■ ■ ■ ■ ■ ■ ■ ■ ■ ■ ■ ■

3.86

40. PLATE 88. $\widetilde{X}$ = 1.70 + .72Y. For score value Y = 8 determine the value of $\widetilde{X}$. For Y = 8, $\widetilde{X}$ = _____ .

■ ■ ■ ■ ■ ■ ■ ■ ■ ■ ■ ■ ■ ■ ■

7.46   [1.70 + .72(8)]

41. Recall that Plate 87 presented the scatter diagram for the data in Plate 88. Recall also that the regression line shown in Plate 87 depicted the regression of ___(symbol) on ___(symbol).

■ ■ ■ ■ ■ ■ ■ ■ ■ ■ ■ ■ ■ ■ ■

Y, X

42. PLATE 88. $\widetilde{X}$ = 1.70 + .72Y. For Y = 3, $\widetilde{X}$ = 3.86; for Y = 8, $\widetilde{X}$ = 7.46. In Plate 87 the regression line shown is for Y on X. Using the above data, you can also draw the regression line representing ___ (symbol) on ___ (symbol).

■ ■ ■ ■ ■ ■ ■ ■ ■ ■ ■ ■ ■ ■ ■

X, Y

Plate 89. Scatter Diagram Indicating Regression Lines

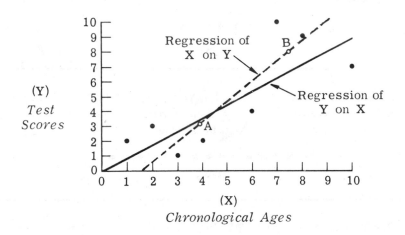

(X)
Chronological Ages

43. In Plate 89, which is for the same data as Plate 88, the dotted line represents the regression of X on Y. This line is determined by the calculations you have just made: for Y = 3, $\widetilde{X}$ = 3.86 (plotted at point A on graph); for Y = 8, $\widetilde{X}$ = 7.46 (plotted at point ___ on graph).

■ ■ ■ ■ ■ ■ ■ ■ ■ ■ ■ ■ ■ ■ ■

B

44. PLATE 89. The dotted line drawn through points A and B represents the regression line of ___ (symbol) on ___ (symbol).

■ ■ ■ ■ ■ ■ ■ ■ ■ ■ ■ ■ ■ ■ ■

X, Y

45. PLATE 89. By use of the regression line of X on Y, you can estimate a child's chronological age if you know his _____ _____ .

■ ■ ■ ■ ■ ■ ■ ■ ■ ■ ■ ■ ■ ■

test score

---

46. PLATE 89. By use of the regression line of Y on X, you can estimate a child's test score if you know his _____ _____ .

■ ■ ■ ■ ■ ■ ■ ■ ■ ■ ■ ■ ■ ■

chronological age

---

47. PLATE 88. $\bar{X}$ = 5.12, $\bar{Y}$ = 4.75. In the scatter diagram presented in Plate 87, locate the point represented by these two mean scores.

■ ■ ■ ■ ■ ■ ■ ■ ■ ■ ■ ■ ■ ■

see Plate 87

---

48. PLATE 89. $\bar{X}$ = 5.12, $\bar{Y}$ = 4.75. The point in the scatter diagram located by the two mean scores represents the point at which the two regression lines _____ .

■ ■ ■ ■ ■ ■ ■ ■ ■ ■ ■ ■ ■ ■

cross

---

49. PLATE 89. The two regression lines cross at the point represented by the means of the two variables. Because all of the dots in this diagram do not lie in a straight line, the correlation between the two variables is _____ _____ (perfect/less than perfect).

■ ■ ■ ■ ■ ■ ■ ■ ■ ■ ■ ■ ■ ■

less than perfect

50. If the correlation between two variables is less than perfect, you can draw two regression lines. However, if the correlation is perfect (1.00 or −1.00), the two regression lines will be _____ (the same/different).

■ ■ ■ ■ ■ ■ ■ ■ ■ ■ ■ ■ ■

the same

EXERCISES

1. What is a regression line?

2. Using Formula 34, calculate the regression coefficient for the regression of Y on X for the following set of data.

Student	Inference Test Scores (X)	Social Studies Test Scores (Y)
A	11	4
B	12	2
C	13	1
D	14	3
E	14	5
F	15	6
G	16	10
H	17	4
I	18	9
J	20	8

3. Using Formulas 33 through 35, predict the social studies test score for a student who receives an inference test score of 13. For one who receives an inference test score of 19.

4. Draw a scatter diagram of the data presented in Exercise 2. Using the values obtained in Exercise 3, draw in this scatter diagram the regression line of Y on X.

5. Using Formula 37, calculate the regression coefficient for the regression of X on Y for the data in Exercise 2.

6. Using Formulas 36 through 38, predict the inference test score for a student who receives a social studies test score of 2. For one who receives a social studies test score of 7.

7. In the scatter diagram drawn for Exercise 4, draw the regression line of X on Y.

8. Identify the symbols $\tilde{Y}, b_{yx}, \tilde{X}, b_{xy}$.

SPEARMAN'S RANK ORDER CORRELATION

In each statistical technique presented so far, the researcher has had to assume that the raw scores were normally distributed in the population from which the sample or samples were selected. If he is unable to make this assumption he must use statistical techniques which do not require such an assumption. Thus, the Pearson product-moment correlation coefficient presented in Set 21 is an appropriate statistical technique for correlation analysis only when the scores in the population from which the samples have been drawn are normally distributed. If this assumption cannot be made, the use of the Pearson product-moment correlation technique is inappropriate, and the evaluation of its significance may be erroneous.

A correlation technique not concerned with the actual value of the raw scores or their distribution in the population is Spearman's *rank order correlation*, or *rho* (ρ). This is an appropriate statistical technique for determining the degree of correlation between two variables when you cannot make the assumption that they come from normally distributed populations. As will be discussed in this set, the rank order correlation does not take into account the value of the raw scores involved; it merely is concerned with the placement of each score in relation to the others in the distribution. This set will present methods for ranking scores and computing the rank order correlation coefficient.

The method for evaluating the significance of the rank order correlation coefficient using Table 5 will be presented, as well as the basis upon which the null hypothesis is accepted or rejected.

SPECIFIC OBJECTIVES FOR SET 23

At the conclusion of this set you will be able to:

(1) rank order a set of scores.

(2) determine the ranks to be assigned to tied scores.

(3) use Formula 39 to calculate the rank order correlation coefficient (rho).

(4) use Table 5 to evaluate the significance of rho.

(5) identify the symbol ρ.

1. If you have a set of scores on a variable that have been arranged in order of magnitude, they are said to be *ranked*. Any set of scores can be _____ .

■ ■ ■ ■ ■ ■ ■ ■ ■ ■ ■ ■ ■ ■

ranked

---

2. If the largest score in a set of ten scores is assigned a rank of 1, the second largest score is assigned a rank of 2, and so on, then the smallest score is assigned a rank of ____ .

■ ■ ■ ■ ■ ■ ■ ■ ■ ■ ■ ■ ■ ■

10

---

3. The term *rank ordered* is used when you place the scores in order of magnitude and assign _____ to them.

■ ■ ■ ■ ■ ■ ■ ■ ■ ■ ■ ■ ■ ■

ranks

Plate 90.

Raw Scores	Ranks
47	1
39	2
38	3
35	4
31	5
29	6
27	7
$N = 7$	

4. This plate presents a set of scores that have been rank ordered. The first column presents the raw scores and the second presents their _____ .

■ ■ ■ ■ ■ ■ ■ ■ ■ ■ ■ ■ ■ ■

ranks

---

5. PLATE 90. Score 27 receives the rank of 7 because it is the smallest score in the set. The largest score in the set is _____ , and it receives a rank of ___ .

■ ■ ■ ■ ■ ■ ■ ■ ■ ■ ■ ■ ■ ■ ■

47, 1

6. PLATE 90. Note that the raw scores are presented in descending order, whereas their ranks are in _____ order.

■ ■ ■ ■ ■ ■ ■ ■ ■ ■ ■ ■ ■ ■ ■

ascending

7. PLATE 90. Although not necessary, it is customary for ranks to be assigned in this manner, with the largest score receiving the rank of ___ .

■ ■ ■ ■ ■ ■ ■ ■ ■ ■ ■ ■ ■ ■ ■

1

Plate 91.

Raw Scores	Ranks
40	1
39	2.5
39	2.5
37	4
20	5
19	6
18	8
18	8
18	8
12	10
10	11
$N = 11$	

8. In this plate, there are _____ raw scores that have been rank ordered.

■ ■ ■ ■ ■ ■ ■ ■ ■ ■ ■ ■ ■ ■ ■

11

9. PLATE 91. Notice that there are two scores of 39 in this plate. These two scores are occupying rank positions of 2 and ___ .

■ ■ ■ ■ ■ ■ ■ ■ ■ ■ ■ ■ ■ ■

3

---

10. PLATE 91. Because the two scores of 39 are "tied" for the second and third ranks, they are both assigned the average of these two ranks, which is

___ .

■ ■ ■ ■ ■ ■ ■ ■ ■ ■ ■ ■ ■ ■

2.5

---

11. Whenever there are "ties" among the raw scores, they are assigned the average of the rank positions that they occupy. In Plate 91, the three raw scores of 18 occupy rank positions of ___ , ___ , and ___ .

■ ■ ■ ■ ■ ■ ■ ■ ■ ■ ■ ■ ■ ■

7, 8, 9

---

12. PLATE 91. The three raw scores of 18 occupy rank positions of 7, 8, and 9. Therefore, they are all three assigned the average of these ranks, which is ___ .

■ ■ ■ ■ ■ ■ ■ ■ ■ ■ ■ ■ ■ ■

8

---

13. Below is a set of raw scores. Place them in rank order and assign ranks to them.

12, 9, 4, 10, 15, 7, 6, 9

■ ■ ■ ■ ■ ■ ■ ■ ■ ■ ■ ■ ■ ■

Plate 92.

Raw Scores	Ranks
15	1
12	2
10	3
9	4.5
9	4.5
7	6
6	7
4	8
$N = 8$	

---

14. PLATE 92. A rank of 4.5 is assigned to both raw scores of 9 because they occupy the rank positions of ___ and ___ in the rank order of the raw scores.

■ ■ ■ ■ ■ ■ ■ ■ ■ ■ ■ ■ ■

4, 5

---

15. If you have two sets of scores for a sample that have been rank ordered, you may compute a correlation between the _____ orders.

■ ■ ■ ■ ■ ■ ■ ■ ■ ■ ■ ■ ■

rank

Plate 93.

Individual	Raw Score X	Raw Score Y	Ranks X	Ranks Y	D	D^2
A	15	9	1	3	−2	4
B	14	11	2	2	0	0
C	12	15	3	1	2	4
D	11	8	4	4	0	0
E	7	4	5.5	5	0.5	0.25
F	7	3	5.5	6	−0.5	0.25
G	4	1	7	7	0	0
$N = 7$						$\Sigma D^2 = 8.5$

16. Plate 93 presents two sets of scores for a sample, each of which has been rank ordered. The first and second columns present the raw scores for the two variables, and the ranks of the raw scores are presented in the _____ and _____ columns.

■ ■ ■ ■ ■ ■ ■ ■ ■ ■ ■ ■ ■ ■ ■

third, fourth

---

17. PLATE 93. For individual C, the rank of his X score is 3 because it is the third largest score for the X variable. The rank of his Y score is ___because it is the _____ score for the Y variable.

■ ■ ■ ■ ■ ■ ■ ■ ■ ■ ■ ■ ■ ■ ■

1, largest

---

18. PLATE 93. Individuals E and F both have raw scores of 7 on the X variable. The ranks of both individuals E and F are 5.5 on this variable because their scores occupy the___ and ___rank positions in the rank order.

■ ■ ■ ■ ■ ■ ■ ■ ■ ■ ■ ■ ■ ■ ■

5, 6

Formula 39. Calculation of Spearman's Rank Order Correlation Coefficient (rho)

$$\rho = 1 - \frac{6\Sigma D^2}{N(N^2 - 1)}$$

in which D = difference between a pair of ranks
N = number of pairs of ranks

19. FORMULA 39. This formula is for the calculation of Spearman's rank order correlation coefficient (rho), the symbol for which is ___ (symbol).

■ ■ ■ ■ ■ ■ ■ ■ ■ ■ ■ ■ ■ ■ ■

ρ

20. FORMULA 39. The symbol ρ is the Greek letter, rho, which is pronounced "row." This is the formula used to compute the correlation coefficient of two sets of _____ (scores/ranks).

■ ■ ■ ■ ■ ■ ■ ■ ■ ■ ■ ■ ■ ■

ranks

21. FORMULA 39. The symbol D in the formula for ρ is used to indicate the _____ between the ranks of an individual.

■ ■ ■ ■ ■ ■ ■ ■ ■ ■ ■ ■ ■ ■

difference

22. FORMULA 39. The symbols $6\Sigma D^2$ indicate that you square the difference between the ranks of the individuals, sum all the squares, and then multiply this sum by___.

■ ■ ■ ■ ■ ■ ■ ■ ■ ■ ■ ■ ■ ■

6

23. FORMULA 39. In this formula, the numbers 1 and 6 are "constants" that are always used when calculating ___(symbol).

■ ■ ■ ■ ■ ■ ■ ■ ■ ■ ■ ■ ■ ■

ρ

24. PLATE 93. In addition to presenting the raw scores and their ranks, this plate also presents the difference between the ranks (D) and the square of the rank differences (D^2). For individual A, $D =$___ and $D^2 = $___.

■ ■ ■ ■ ■ ■ ■ ■ ■ ■ ■ ■ ■ ■

$-2, 4$

25. PLATE 93. In this plate, ΣD^2 = _____ and $N = $___.

■ ■ ■ ■ ■ ■ ■ ■ ■ ■ ■ ■ ■ ■

8.5, 7

26. PLATE 93. $\Sigma D^2 = 8.5$, $N = 7$. To obtain the rank order correlation coefficient, substitute the above values for the symbols in Formula 39.

$$\rho = 1 - \frac{6\Sigma D^2}{N(N^2 - 1)} = 1 - \frac{\underline{}}{\underline{}}$$

■ ■ ■ ■ ■ ■ ■ ■ ■ ■ ■ ■ ■ ■

$$\frac{6(8.5)}{7(7^2 - 1)}$$

---

27. PLATE 93. $\Sigma D^2 = 8.5$, $N = 7$.

$$\rho = 1 - \frac{6(8.5)}{7(7^2 - 1)} = 1 - \frac{\underline{}}{\underline{}} = 1 - \underline{} = \underline{} \text{ (to two decimal places).}$$

■ ■ ■ ■ ■ ■ ■ ■ ■ ■ ■ ■ ■ ■

$$1 - \frac{51}{336} = 1 - .15 = .85$$

---

28. PLATE 93. The rank order correlation coefficient for this set of ranks is $\rho = .85$. Table 5 presents the values of ρ required for significance at the _____ and _____levels.

■ ■ ■ ■ ■ ■ ■ ■ ■ ■ ■ ■ ■ ■

.05, .01

---

29. Notice that the first column in Table 4 for r is headed df, whereas the first column in Table 5 for ρ is headed ___ (symbol).

■ ■ ■ ■ ■ ■ ■ ■ ■ ■ ■ ■ ■ ■

N

---

30. Because the ρ correlation coefficient is computed on ranks rather than raw scores, you do not need to determine df. Table 5 is entered directly by using N, which is the number of _____ of ranks in the correlation.

■ ■ ■ ■ ■ ■ ■ ■ ■ ■ ■ ■ ■ ■

pairs

---

31. PLATE 93. $\rho = .85$. For the data in this plate, $N =$ ___ .

■ ■ ■ ■ ■ ■ ■ ■ ■ ■ ■ ■ ■ ■

7

---

32. PLATE 93. $\rho = .85$, $N = 7$. Use Table 5 to determine the value of ρ required for significance. $P = .05$, $\rho =$ _____ . $P = .01$, $\rho =$ _____ .

■ ■ ■ ■ ■ ■ ■ ■ ■ ■ ■ ■ ■ ■

.786, 929

---

33. PLATE 93. $\rho = .85$, $N = 7$. FROM TABLE 5: for $P = .05$, $\rho = .786$; for $P = .01$, $\rho = .929$. The obtained ρ of .85 is _____ (larger/smaller) than required at the .05 level of significance.

■ ■ ■ ■ ■ ■ ■ ■ ■ ■ ■ ■ ■ ■

larger

---

34. The null hypothesis states that there is ____ correlation between the ranks of the X and Y variables.

■ ■ ■ ■ ■ ■ ■ ■ ■ ■ ■ ■ ■ ■

no

---

35. PLATE 93. Because the obtained ρ of .85 is larger than is required for significance at the .05 level, you _____ (accept/reject) the null hypothesis.

■ ■ ■ ■ ■ ■ ■ ■ ■ ■ ■ ■ ■ ■

reject

36. Table 5 is for two-tailed significance tests. Therefore, the values in this table apply to both positive and _____ rank order correlations.

■ ■ ■ ■ ■ ■ ■ ■ ■ ■ ■ ■ ■

negative

EXERCISES

1. A group of soldiers were given instruction in marksmanship. A study was conducted to determine if the number of hours spent in practice shooting was related to proficiency. Below are presented the data for twelve soldiers. Use Formula 39 to compute the rank order correlation for these data.

Soldier	Hours of Practice	Proficiency Rating
A	24	75
B	23	83
C	19.5	98
D	18	80
E	17	74
F	16.5	69
G	16	71
H	15	68
I	14.5	59
J	14	62
K	13.5	70
L	10	54

2. Using Table 5, determine if the rho computed in Exercise 1 is significant. Use $P = .01$ as your designated level for significance. Would you accept or reject the null hypothesis?

3. A teacher was interested in knowing the extent to which his evaluation of his children's cooperativeness was related to their evaluation of themselves. He rated each child on cooperativeness, using a scale ranging from 1, for very cooperative, to 10, for uncooperative. Each child also rated himself, using the same scale. Below are the data obtained for ten children. Compute the rank order correlation for these data, using Formula 39.

Child	Teacher's Rating	Child's Self-rating	Child	Teacher's Rating	Child's Self-rating
A	5	4	F	8	2
B	10	2	G	4	3
C	1	3	H	4	5
D	6	1	I	6	5
E	2	3	J	6	6

4. Using Table 5, determine if the rho computed in Exercise 3 is significant. Use $P = .05$ as your designated level for significance. Would you accept or reject the null hypothesis?

5. Identify the symbol ρ (rho).

311

CHI SQUARE:
SINGLE SAMPLE

Previous sets of this book have analyzed data in the form of raw scores or ranks. Another type of data with which the researcher is concerned is *frequency data*. These data are not in the form of raw scores or ranks, but they indicate the *frequency* with which events occur—for instance, the number of boys preferring one soft drink over another or the number of college students attending a concert. Any number which denotes how many times an occurrence takes place is called frequency data.

Frequency data can be divided into various categories, and the differences within the categories can be analyzed. One statistical technique appropriate for examining frequency data is called the *chi square* technique (χ^2).

This set presents the method for calculating chi square from the frequency data obtained from one sample and the method of evaluating the significance of the chi square.

SPECIFIC OBJECTIVES OF SET 24

At the conclusion of this set you will be able to:

(1) use Formula 40 to compute the chi square from frequency data divided into two cells.

(2) use Formula 41 to compute chi square from frequency data divided into more than two cells.

(3) use Table 6 to determine the level of significance of chi square.

(4) identify the symbols O, E, and χ^2

1. Thus far you have applied statistical techniques to raw scores or ranks of scores. You may also apply statistical tests to the frequencies of individuals when they are categorized in various ways.

Plate 94. Question: "Are you in favor of capital punishment?"

	f
Yes	20
No	60
	$N = 80$

The frequency of people answering "yes" to the question in Plate 94 is ____ .

■ ■ ■ ■ ■ ■ ■ ■ ■ ■ ■ ■ ■

20

---

2. PLATE 94. Out of a sample of eighty people who were asked this question, _____ answered "yes" and _____ answered "no."

■ ■ ■ ■ ■ ■ ■ ■ ■ ■ ■ ■ ■

twenty, sixty

---

3. PLATE 94. In this plate, the numbers 20 and 60 are _____ (frequencies/scores).

■ ■ ■ ■ ■ ■ ■ ■ ■ ■ ■ ■ ■

frequencies

---

4. PLATE 94. In this example, "yes" and "no" are called *categories*. Suppose you wish to answer this question: "Are the frequencies of individuals in the two _____ significantly different?"

■ ■ ■ ■ ■ ■ ■ ■ ■ ■ ■ ■ ■

categories

---

5. **PLATE 94.** The null hypothesis to be tested is: There is ____ difference in the population between the number of people giving "yes" and "no" answers.

■ ■ ■ ■ ■ ■ ■ ■ ■ ■ ■ ■ ■ ■

no

---

6. **PLATE 94.** If there is no difference between the frequency of individuals giving "yes" and "no" answers, you would expect that ____ % of the individuals would fall into each category.

■ ■ ■ ■ ■ ■ ■ ■ ■ ■ ■ ■ ■ ■

50

---

7. **PLATE 94.** If you expect 50% of the individuals to answer "yes," then the expected frequency of "yes" answers for $N = 80$ is ____.

■ ■ ■ ■ ■ ■ ■ ■ ■ ■ ■ ■ ■ ■

40

Plate 95.

	O	E	$|O - E| - .5$	$(|O - E| - .5)^2$	$\dfrac{(|O - E| - .5)^2}{E}$
Yes	20	40	19.5	380.25	9.5
No	60	40	19.5	380.25	9.5

8. The frequencies obtained in each of the two categories are commonly called the *observed* frequencies. In this plate, the observed frequencies are ____ for the "yes" category, and ____ for the "no" category.

■ ■ ■ ■ ■ ■ ■ ■ ■ ■ ■ ■ ■ ■

20, 60

---

9. PLATE 95. This plate presents the same data as Plate 94. The column marked O presents the observed frequencies, and the column marked E presents the _____ frequencies.

■ ■ ■ ■ ■ ■ ■ ■ ■ ■ ■ ■ ■ ■

expected

Formula 40. Calculation of the Chi Square when $df = 1$

$$\chi^2 = \sum \frac{(|O - E| - .5)^2}{E}$$

in which O = observed frequency
E = expected frequency

The subtraction of .5 from each $|O - E|$ represents the Yates correction for continuity.

---

10. The statistical test applied to determine if the observed frequencies are significantly different from the expected frequencies is presented in Formula 40. This is called a _____ _____ test.

■ ■ ■ ■ ■ ■ ■ ■ ■ ■ ■ ■ ■ ■

chi square

---

11. FORMULA 40. The Greek symbol that represents the chi square is _____ (symbol).

■ ■ ■ ■ ■ ■ ■ ■ ■ ■ ■ ■ ■ ■

χ^2

---

12. FORMULA 40. In the formula, the symbols $|O - E| - .5$ indicate that you obtain the absolute difference between the O and E for each category, and then subtract ____ from this difference.

■ ■ ■ ■ ■ ■ ■ ■ ■ ■ ■ ■ ■ ■

.5

---

13. FORMULA 40. For each category, you subtract .5 from the _____ difference between O and E.

■ ■ ■ ■ ■ ■ ■ ■ ■ ■ ■ ■ ■ ■ ■

absolute

---

14. FORMULA 40. By subtracting .5 from the absolute difference between O and E, you are reducing this difference. This is called the Yates correction for

_____.

■ ■ ■ ■ ■ ■ ■ ■ ■ ■ ■ ■ ■ ■ ■

continuity

---

15. FORMULA 40. The symbols $(|O - E| - .5)^2/E$ indicate that you obtain the absolute difference between O and E, reduce it by .5, square it, and divide by ___ (symbol).

■ ■ ■ ■ ■ ■ ■ ■ ■ ■ ■ ■ ■ ■

E

---

16. FORMULA 40. The symbol Σ in this formula indicates that in order to obtain χ^2, you _____ the quotients for all the categories.

■ ■ ■ ■ ■ ■ ■ ■ ■ ■ ■ ■ ■ ■

sum

---

17. Plate 95 presents the application of the χ^2 test. The third column of this plate presents $|O - E| - .5$ for each of the categories. The fourth column presents _____ for each of the categories.

■ ■ ■ ■ ■ ■ ■ ■ ■ ■ ■ ■ ■ ■

$(|O - E| - .5)^2$

---

18. PLATE 95. For the "yes" category: $O = 20$, $E = 40$. $|O - E| - .5 =$ _____ ; $(|O - E| - .5)^2 =$ _____ .

■ ■ ■ ■ ■ ■ ■ ■ ■ ■ ■ ■ ■ ■

19.5, 380.25

19. PLATE 95. The fifth column presents $(|O - E| - .5)^2/E$. This is obtained by dividing $(|O - E| - .5)^2$ by E for each of the categories. For the "yes" category: $(|O - E| - .5)^2/E =$ _____ .

■ ■ ■ ■ ■ ■ ■ ■ ■ ■ ■ ■ ■ ■

9.5

20. PLATE 95. The same procedure is followed for the "no" category. For the "no" category: $(|O - E| - .5)^2/E =$ _____ .

■ ■ ■ ■ ■ ■ ■ ■ ■ ■ ■ ■ ■ ■

9.5

21. FORMULA 40. This formula indicates that, in order to obtain χ^2, you must _____ $(|O - E| - .5)^2/E$ of the categories.

■ ■ ■ ■ ■ ■ ■ ■ ■ ■ ■ ■ ■ ■

sum

22. PLATE 95. For these data:

$$\chi^2 = \sum \frac{(|O - E| - .5)^2}{E} = 9.5 + 9.5 = \underline{\hspace{1cm}}$$

■ ■ ■ ■ ■ ■ ■ ■ ■ ■ ■ ■ ■ ■

19

23. FORMULA 40. The term *cells* is used to indicate the classifications into which the observed frequencies are divided. In Plate 95 there are _____ cells.

■ ■ ■ ■ ■ ■ ■ ■ ■ ■ ■ ■ ■ ■

two

24. PLATE 95. There are two cells into which the observed frequencies are divided. There is the "yes" cell and the "____" cell.

■ ■ ■ ■ ■ ■ ■ ■ ■ ■ ■ ■ ■ ■

no

---

25. To determine if the value of the χ^2 indicates that the frequencies in the two cells are significantly different, you must first determine the _____ of freedom.

■ ■ ■ ■ ■ ■ ■ ■ ■ ■ ■ ■ ■ ■

degrees

---

26. Recall that, when dealing with raw scores, the *df* is determined by how many raw scores are free to vary. When dealing with frequencies, as in χ^2, the *df* is determined by how many cells have frequencies which are free to_____.

■ ■ ■ ■ ■ ■ ■ ■ ■ ■ ■ ■ ■

vary

---

27. The degrees of freedom for χ^2 is determined by the number of _____ which have frequencies free to vary.

■ ■ ■ ■ ■ ■ ■ ■ ■ ■ ■ ■ ■ ■

cells

---

28. PLATE 95. There are two cells for these data. If, for the "yes" cell, the frequency is known, then the frequency for the "no" cell is "fixed" because the sum of the frequencies must equal 80. Thus, for these data, one cell is not_____ to _____.

■ ■ ■ ■ ■ ■ ■ ■ ■ ■ ■ ■ ■

free, vary

29. The *df* associated with χ^2 is the number of cells that are free to vary. When your data are divided into two cells, the *f* of one cell is free to vary; but once it is determined, the *f* of the other cell is fixed (that is, *not* free to vary). For the data in Plate 95, *df* = ____ .

■ ■ ■ ■ ■ ■ ■ ■ ■ ■ ■ ■ ■ ■

1 $(2-1)$

---

30. PLATE 95. For these data, *df* = 1 because only one cell has frequencies free to _____ .

■ ■ ■ ■ ■ ■ ■ ■ ■ ■ ■ ■ ■ ■

vary

---

31. Table 6 presents the values of χ^2 needed for the .05 and .01 levels of significance for the differing *df*'s. For *df* = 1, the value of χ^2 required for significance at the .05 level is_____ and at the .01 level is_____ .

■ ■ ■ ■ ■ ■ ■ ■ ■ ■ ■ ■ ■ ■

3.84, 6.64

---

32. PLATE 95. χ^2 = 19, *df* = 1. FROM TABLE 6: For *P* = .05, χ^2 = 3.84; for *P* = .01, χ^2 = 6.64. Because the obtained χ^2 of 19 is larger than either of the two χ^2 values listed in Table 6, you conclude that the difference between the frequencies in the two categories is significant beyond the _____ level.

■ ■ ■ ■ ■ ■ ■ ■ ■ ■ ■ ■ ■ ■

.01

---

33. PLATE 95. χ^2 = 19, *df* = 1. Because the obtained χ^2 of 19 is larger than is required for significance at the .01 level, you _____ (accept/reject) the hypothesis that there is no difference between the frequencies in the "yes" and "no" categories.

■ ■ ■ ■ ■ ■ ■ ■ ■ ■ ■ ■ ■ ■

reject

319

34. PLATE 95. On the basis of the rejection of the null hypothesis at the .01 level, you conclude that in the population from which this sample was selected, most people _____(approve/disapprove) of capital punishment.

■ ■ ■ ■ ■ ■ ■ ■ ■ ■ ■ ■ ■ ■

disapprove

---

35. In cases where you have more than one degree of freedom, it is not necessary to apply the _____correction for _____.

■ ■ ■ ■ ■ ■ ■ ■ ■ ■ ■ ■ ■ ■

Yates, continuity

---

36. The Yates correction for continuity need only be applied in cases where you have only_____degree of freedom.

■ ■ ■ ■ ■ ■ ■ ■ ■ ■ ■ ■ ■ ■

one

Formula 41. Calculation of the Chi Square when *df* Is Larger than 1

$$\chi^2 = \sum \frac{(O - E)^2}{E}$$

37. FORMULA 41. This formula is for the calculation of χ^2 when you have more than one degree of freedom. Notice that in this formula you _____ (do/do not) subtract .5 from the absolute difference between O and E.

■ ■ ■ ■ ■ ■ ■ ■ ■ ■ ■ ■ ■ ■

do not

Plate 96. Question: "Which kind of coffee do you prefer?"

	O	E	$O - E$	$(O - E)^2$	$\dfrac{(O - E)^2}{E}$
Brand A	35				
Brand B	28				
Brand C	30				
Brand D	35				
	$N = 128$				

38. Chi square tests need not be limited in use to only two cells. Plate 96 presents the frequencies of individuals' preferences for four brands of coffee. In this sample, $N =$ _____.

■ ■ ■ ■ ■ ■ ■ ■ ■ ■ ■ ■ ■ ■

128

---

39. PLATE 96. The chi square test for these data will have more than one degree of freedom. Therefore, you will use Formula_____(40/41) in computing the chi square.

■ ■ ■ ■ ■ ■ ■ ■ ■ ■ ■ ■ ■ ■

41

---

40. PLATE 96. $N = 128$. In testing the null hypothesis, you would expect that the frequencies are equally divided among the four cells. This means that your expected frequency for each cell in this plate is _____ .

■ ■ ■ ■ ■ ■ ■ ■ ■ ■ ■ ■ ■ ■

32

---

41. PLATE 96. Your expected frequency for each brand is 32. Use Formula 41. For Brand A compute the following:

$O - E =$ _____ $(O - E)^2 =$ _____ $\dfrac{(O - E)^2}{E} =$ _____

■ ■ ■ ■ ■ ■ ■ ■ ■ ■ ■ ■ ■ ■

3, 9, .28

42. **PLATE 96.** For Brand A: $E = 32$, $O - E = 3$, $(O - E)^2 = 9$, $(O - E)^2/E = .28$. Determine $(O - E)^2/E$ for the other three brands of coffee. Brand B _____ , Brand C _____ , Brand D _____ (all to two decimal places).

■ ■ ■ ■ ■ ■ ■ ■ ■ ■ ■ ■ ■ ■

.50, .12, .28

---

43. **PLATE 96.** $(O - E)^2/E$. Brand A = .28, Brand B = .50, Brand C = .12, Brand D = .28.

FORMULA 41. In order to determine χ^2 you must _____ the above values.

■ ■ ■ ■ ■ ■ ■ ■ ■ ■ ■ ■ ■ ■

sum

---

44. **PLATE 96.** The chi square for these data is:

$$\chi^2 = .28 + .50 + .12 + .28 = \underline{\hspace{2cm}} .$$

■ ■ ■ ■ ■ ■ ■ ■ ■ ■ ■ ■ ■ ■

1.18

---

45. **PLATE 96.** $\chi^2 = 1.18$. The frequencies in this plate are categorized into _____ cells.

■ ■ ■ ■ ■ ■ ■ ■ ■ ■ ■ ■ ■ ■

four

---

46. **PLATE 96.** This plate has four cells. If the frequencies contained in three of the cells are known, the frequency of the fourth cell is fixed. Therefore, for these data, the $df = \underline{\hspace{1cm}} .$

■ ■ ■ ■ ■ ■ ■ ■ ■ ■ ■ ■ ■ ■

3

47. PLATE 96. For these data, $df = 3$, because only three of the four cells have frequencies free to vary. Once these three frequencies are known, the frequency of the fourth cell is fixed because the sum of all the frequencies must equal ___ (symbol).

■ ■ ■ ■ ■ ■ ■ ■ ■ ■ ■ ■ ■ ■

N

48. PLATE 96. $\chi^2 = 1.18$, $df = 3$. From Table 6, for $df = 3$ the value of χ^2 required to be significant at the .05 level is _____ , and at the .01 level is _____ .

■ ■ ■ ■ ■ ■ ■ ■ ■ ■ ■ ■ ■ ■

7.82, 11.34

49. PLATE 96. $\chi^2 = 1.18$, $df = 3$. FROM TABLE 6: For $P = .05$, $\chi^2 = 7.82$; for $P = .01$, $\chi^2 = 11.34$. Because the obtained χ^2 of 1.18 is not as large as either of the two χ^2 values listed in Table 6, you conclude that the observed differences among the four cells are _____ (significant/non-significant).

■ ■ ■ ■ ■ ■ ■ ■ ■ ■ ■ ■ ■ ■

non-significant

50. PLATE 96. $\chi^2 = 1.18$, $df = 3$. Because the obtained χ^2 is less than is required for significance, you _____ (accept/reject) the null hypothesis that there is no significant difference among the frequencies in the four cells.

■ ■ ■ ■ ■ ■ ■ ■ ■ ■ ■ ■ ■ ■

accept

51. PLATE 96. By accepting the null hypothesis, you conclude that the observed differences in the frequencies among the cells is due to _____ _____ .

■ ■ ■ ■ ■ ■ ■ ■ ■ ■ ■ ■ ■ ■

sampling error

EXERCISES

1. A group of college students were shown a series of pictures illustrating a new clothing style. After viewing the pictures, each student was asked if he approved or disapproved of the style. Below are the data received from fifty-four students.

Approve	35
Disapprove	19

    Compute the chi square for the above data. Determine whether to use Formula 40 or 41.

2. Using $P = .05$ as your designated level for significance, use Table 6 to evaluate the chi square obtained in Exercise 1. State the null hypothesis being tested. On the basis of the chi square test, do you accept or reject the null hypothesis?

3. A researcher wished to examine children's preferences among four types of transportation. A sample of ninety children was randomly selected and asked which type they preferred. The following data were obtained:

Automobile	10
Bus	13
Train	27
Airplane	40

    Compute the chi square for the above data. Determine whether to use Formula 40 or 41.

4. Using $P = .05$ as your acceptable level for significance, use Table 6 to evaluate the chi square obtained in Exercise 3. State the null hypothesis being tested. On the basis of the chi square test, do you accept or reject the null hypothesis?

5. A manager of a large manufacturing firm wished to know if the installation of an employees' lounge had increased or decreased the productivity of his employees. He selected a sample of forty employees and obtained the following data on the change in their productivity.

Productivity increased	21
No change in productivity	11
Productivity decreased	8

    Compute the chi square for the above data.

6. Using $P = .05$ as your acceptable level of significance, evaluate the chi square obtained in Exercise 5. Would you accept or reject the null hypothesis?

7. A pharmaceutical firm developed a new medicine to alleviate discomfort of the common cold. In order to test its acceptance, a sample of three hundred people with colds was given the medicine. They were then asked if they thought the medicine was effective. Here are the data obtained.

Yes	171
No	129

Compute the chi square for the above data.

8. Using $P = .01$ as your acceptable level for significance, evaluate the chi square obtained in Exercise 7. Would you accept or reject the null hypothesis?

9. Identify the symbol χ^2.

CHI SQUARE:
MULTIPLE SAMPLES

The previous set presented the chi square test for frequencies obtained from one sample. Frequently, the researcher obtains data from more than one sample, and these data are divided into a number of categories for each sample. The chi square technique can also be applied when there are multiple samples providing data. This set presents the use of the chi square technique to test the differences among the frequencies for various samples categorized separately. Methods are given for calculating expected frequencies and determining the significance level of chi square, and a formula is presented which determines the degrees of freedom associated with the chi square.

SPECIFIC OBJECTIVES OF SET 25

At the conclusion of the set you will be able to:

(1) use Formula 42 to determine expected frequencies for the cells, using the observed frequencies obtained from several samples.

(2) use Formula 41 to calculate the chi square for multiple samples.

(3) Determine the degrees of freedom associated with chi square for multiple samples, and Table 6 to determine its level of significance.

1. Thus far you have been dealing with one group of individuals which has been divided into cells according to categories of responses.

Plate 97. Observed Frequencies. Question: "Do you enjoy classical music?"

	Yes	Undecided	No	N_{row}
Boys	A 46	B 10	C 30	86
Girls	D 20	E 18	F 50	88
N_{col}	66	28	80	$N_{total} = 174$

Plate 97 has three categories of responses for _____ groups of individuals.

■ ■ ■ ■ ■ ■ ■ ■ ■ ■ ■ ■ ■ ■

two

2. PLATE 97. A group of boys and a group of girls were asked this question. The total number of children in the two groups was $N_{total} =$ _____ .

■ ■ ■ ■ ■ ■ ■ ■ ■ ■ ■ ■ ■ ■

174

3. PLATE 97. The number of boys is 86, indicated at the right of the plate under the heading N_{row}, which is the sum of the three cells in the top row. (Cells A, B, and C.) The number of girls is _____, which is the sum of the three cells in the _____ row. (Cells D, E, and F.)

■ ■ ■ ■ ■ ■ ■ ■ ■ ■ ■ ■ ■ ■

88, bottom

4. PLATE 97. Of the 174 children, 66 gave "yes" responses (total of the first column); 28 gave "undecided" responses (total of the second column); and _____ gave "no" responses (total of the third column).

■ ■ ■ ■ ■ ■ ■ ■ ■ ■ ■ ■ ■ ■

80

5. PLATE 97. Of the 86 boys, 46 answered "yes" (Cell A); 10 answered "undecided" (Cell B); and _____ answered "no" (Cell C).

■ ■ ■ ■ ■ ■ ■ ■ ■ ■ ■ ■ ■ ■

30

---

6. PLATE 97. The null hypothesis for these data is: "There is no difference between boys and girls in the type of response given to this question." Because there are three categories of responses and two groups of individuals, the frequencies are divided into _____ cells.

■ ■ ■ ■ ■ ■ ■ ■ ■ ■ ■ ■ ■ ■

six

---

7. PLATE 97. A chi square test can be performed to determine whether the frequencies in the boys' cells differ significantly from the _____ in the _____ cells.

■ ■ ■ ■ ■ ■ ■ ■ ■ ■ ■ ■ ■ ■

frequencies, girls'

---

8. Plate 97 presents the observed frequencies for each of the six cells. In order to perform a χ^2 test, you must determine the _____ frequency for each cell.

■ ■ ■ ■ ■ ■ ■ ■ ■ ■ ■ ■ ■ ■

expected

---

9. PLATE 97. To determine the expected frequency (E) for Cell A, first determine the N of the row in which Cell A is located. N_{row} = ____ .

■ ■ ■ ■ ■ ■ ■ ■ ■ ■ ■ ■ ■ ■

86

---

10. PLATE 97. For Cell A, N_{row} = 86. Next determine the N of the column in which Cell A is located. $N_{col} = \underline{\quad}$.

■ ■ ■ ■ ■ ■ ■ ■ ■ ■ ■ ■ ■ ■

66

Formula 42. Calculation of the Expected Frequency (E) of a Cell

$$E = \frac{(N_{row})(N_{col})}{N_{total}}$$

Degrees of Freedom: (number of rows − 1)(number of columns − 1)

11. FORMULA 42. This formula can be used to determine the expected frequency for each cell.

PLATE 97. To determine the E for Cell A, multiply N_{row} (containing Cell A) by N_{col} (containing Cell A) and divide by _____(symbol).

■ ■ ■ ■ ■ ■ ■ ■ ■ ■ ■ ■ ■ ■

N_{total}

---

12. PLATE 97. For Cell A, N_{row} = 86, N_{col} = 66. Use Formula 42 to determine E for Cell A. $E = \underline{\quad}$.

■ ■ ■ ■ ■ ■ ■ ■ ■ ■ ■ ■ ■ ■

32.6 $\dfrac{(86)(66)}{174}$

---

13. PLATE 97. For Cell E, $N_{row} = \underline{\quad}$ and $N_{col} = \underline{\quad}$.

■ ■ ■ ■ ■ ■ ■ ■ ■ ■ ■ ■ ■ ■

88, 28

---

14. PLATE 97. For Cell E, N_{row} = 88, N_{col} = 28. Use Formula 42 to determine E for Cell E. $E = \underline{\quad}$.

■ ■ ■ ■ ■ ■ ■ ■ ■ ■ ■ ■ ■ ■

14.2 $\dfrac{(88)(28)}{174}$

329

15. For Cell A, $E = 32.6$; for Cell E, $E = 14.2$. PLATE 97. Use Formula 42 to determine the expected frequency (E) for each of the other cells in this plate. Cell B____, Cell C____, Cell D____, Cell F____.

■ ■ ■ ■ ■ ■ ■ ■ ■ ■ ■ ■ ■ ■

Plate 98. Expected Frequencies

A 32.6	B 13.8	C 39.5
D 33.4	E 14.2	F 40.5

16. PLATE 98. This plate presents the expected frequency for each cell, corresponding to the observed frequencies in Plate 97. Thus, for Cell D, $O =$ ____ and $E =$ _____.

■ ■ ■ ■ ■ ■ ■ ■ ■ ■ ■ ■ ■ ■

20, 33.4

17. PLATES 97 and 98. To compute the value of χ^2 for these frequencies, use Formula 41 as you did before. For Cell A, $O = 46$ and $E = 32.6$.

$O - E =$ _____ $(O - E)^2 =$ _____ $\dfrac{(O - E)^2}{E} =$ _____

■ ■ ■ ■ ■ ■ ■ ■ ■ ■ ■ ■ ■ ■

13.4, 179.6, 5.51

18. PLATES 97 and 98. For Cell A, $(O - E)^2/E = 5.51$. Determine this value for each of the other cells. B____, C ____, D____, E ____, F____.

■ ■ ■ ■ ■ ■ ■ ■ ■ ■ ■ ■ ■ ■

1.04, 2.28, 5.37, 1.01, 2.23

19. PLATES 97 and 98. Use Formula 41 to determine χ^2. Sum $(O - E)^2/E$ for all cells. $\chi^2 =$ _____.

■ ■ ■ ■ ■ ■ ■ ■ ■ ■ ■ ■ ■ ■

17.44

$(5.51 + 1.04 + 2.28 + 5.37 + 1.01 + 2.23)$

20. The degrees of freedom associated with the chi square technique, as given in Formula 42, is calculated by:

$df =$ (number of rows $- 1$) (number of columns $- 1$)

For Plate 97, $df =$ ___.

■ ■ ■ ■ ■ ■ ■ ■ ■ ■ ■ ■ ■ ■

2 $(2 - 1)(3 - 1)$

21. PLATE 97. $df = 2$. Thus two cells have frequencies that are free to vary. This means that, for $N_{total} = 174$, if the f of two of the cells are known, the f of the other ___ cells is fixed.

■ ■ ■ ■ ■ ■ ■ ■ ■ ■ ■ ■ ■ ■

4

22. PLATES 97 and 98. $\chi^2 = 17.44$, $df = 2$. From Table 6, for $df = 2$, the value of χ^2 required for significance at the .05 level is _____ , and at the .01 level is _____ .

■ ■ ■ ■ ■ ■ ■ ■ ■ ■ ■ ■ ■ ■

5.99, 9.21

23. PLATES 97 and 98. $\chi^2 = 17.44$, $df = 2$. FROM TABLE 6: For $P = .05$, $\chi^2 = 5.99$; for $P = .01$, $\chi^2 = 9.21$. Because the obtained χ^2 of 17.44 is larger than either of the two χ^2 values listed in Table 6, you conclude that the differences among the frequencies in the six cells is significant beyond the _____level.

■ ■ ■ ■ ■ ■ ■ ■ ■ ■ ■ ■ ■ ■

.01

24. **PLATES 97 and 98.** Because the obtained χ^2 is larger than is required for significance at the .01 level, you _____(accept/reject) the hypothesis that there is no difference between boys and girls in the type of responses given to this question.

■ ■ ■ ■ ■ ■ ■ ■ ■ ■ ■ ■ ■ ■

reject

---

25. **PLATES 97 and 98.** On the basis of the rejection of the null hypothesis at the .01 level, you conclude that in the population from which these samples were selected, classical music is preferred more by_____(boys/girls) than by_____(boys/girls).

■ ■ ■ ■ ■ ■ ■ ■ ■ ■ ■ ■ ■ ■

boys, girls

---

26. The chi square test should not be used when the expected frequency (E) in any cell is less than 5. Thus, if, in Plate 98, E for one of the cells had been 4, the use of the χ^2 test would have been_____ _____ (appropriate/inappropriate).

■ ■ ■ ■ ■ ■ ■ ■ ■ ■ ■ ■ ■ ■

inappropriate

---

27. In order to apply the chi square test properly, the E of each cell must be at least ___ .

■ ■ ■ ■ ■ ■ ■ ■ ■ ■ ■ ■ ■ ■

5

---

28. When you have only one degree of freedom, you must use Formula____ (40/41) in computing χ^2.

■ ■ ■ ■ ■ ■ ■ ■ ■ ■ ■ ■ ■ ■

40

29. When you have more than one degree of freedom, it is not necessary to apply the Yates correction for continuity. Therefore, you may use Formula ____ (40/41).

■ ■ ■ ■ ■ ■ ■ ■ ■ ■ ■ ■ ■ ■

41

30. The chi square test may be applied to any number of groups that have been divided into any number of categories. If you have five groups, each of which is divided into seven categories, your chi square test will consist of _____ cells.

■ ■ ■ ■ ■ ■ ■ ■ ■ ■ ■ ■ ■ ■

thirty-five

31. If you have five groups, each of which has seven categories, the degrees of freedom that you have in your chi square test is ____ .

■ ■ ■ ■ ■ ■ ■ ■ ■ ■ ■ ■ ■

24

EXERCISES

1. A physical education instructor wished to determine preferences of boys and girls for three activities. He asked a sample of forty boys and thirty girls to state their preferences. Here are the data he received.

	Basketball	Volleyball	Kickball
Boys	20	5	15
Girls	7	13	10

Compute the chi square for the above data.

2. Using $P = .05$ as your acceptable level for significance, evaluate the chi square obtained in Exercise 1. Would you accept or reject the null hypothesis? If you reject it, what can be said regarding the preferences of boys and girls?

3. A television network wished to determine whether certain types of programs appealed to different nationality groups. An interviewer asked people in four nationality groups to indicate their preference among five types of television programs. Here are the data he obtained.

	French	German	Italian	Chinese
Program type A	20	18	16	14
Program type B	16	14	18	16
Program type C	14	14	12	18
Program type D	20	16	14	16
Program type E	18	20	14	18

Compute the chi square for the above data.

4. Using $P = .05$ as your acceptable level for significance, evaluate the chi square obtained in Exercise 3. Would you accept or reject the null hypothesis? If you reject it, what can be said about the preferences of different nationality groups?

ANSWERS TO EXERCISES

Set 1

1.	X		f		2.	*interval*		f
	15		1			13-15		1
	14		0			10-12		4
	13		0			7-9		5
	12		1			4-6		3
	11		2			1-3		2
	10		1				$N = \overline{15}$	
	9		2					
	8		2					
	7		1		3.	*interval*		f
	6		1			41-44		2
	5		2			37-40		4
	4		0			33-36		6
	3		0			29-32		3
	2		1			25-28		0
	1		1			21-24		2
			$N = \overline{15}$				$N = \overline{17}$	

4. 8.5; 106.5; .5; 35.5.

5. 10.5; .5; 3.5.

6. X = raw score, f = frequency, N = number of scores, i = size of class interval, $\ell\ell$ = real lower limit.

Set 2

1. The *mode* is the most recurring or frequent score or class interval in a frequency distribution.

2. The *median* is the midpoint or center of the frequency distribution. It is the point above which and below which 50% of the raw scores lie. The symbol for median is *Mdn.*

3. (a) Mode = 9, *Mdn* = 8; (b) Mode = 109, *Mdn* = 107.7; (c) Modal interval = 83-86, *Mdn* = 85.17; (d) Modal interval = 10-19, *Mdn* = 22.5.

4. Σ = the sum of, Σf_b = the sum of frequencies below the interval which contains the median, f_w = the frequency within the interval which contains the median.

Set 3

1. The *mean* of a frequency distribution is the arithmetic average of the raw scores. Its symbol is $\bar{X}$.

2. 3, 5.5, 105.5, 82.

3. (a) $\bar{X}$ = 6.36, (b) $\bar{X}$ = 57.24, (c) $\bar{X}$ = 14.79, (d) $\bar{X}$ = 50.88.

4.

X	f	$(X - 50)$	$f(X - 50)$
59	3	9	27
58	4	8	32
57	6	7	42
56	2	6	12
55	2	5	10
$N = 17$			$\Sigma f(X - 50) = 123$

$$\bar{X} = \frac{123}{17} + 50 = 57.24$$

5.

Interval	f	$(X - 40)$	$f(X - 40)$
57-60	2	18.5	37
53-56	5	14.5	72.5
49-52	9	10.5	94.5
45-48	3	6.5	19.5
41-44	2	2.5	5
$N = 21$			$\Sigma f(X - 40) = 228.5$

$$\bar{X} = \frac{228.5}{21} + 40 = 50.88$$

Set 4

1.	X	(X + 5)	f		f(X + 5)
	2	7	2		14
	1	6	3		18
	0	5	5		25
	−1	4	4		16
	−2	3	3		9
	−3	2	3		6
	−4	1	1		1
			$N = 21$		$\Sigma f(X + 5) = 89$

$$\bar{X} = \frac{89}{21} - 5 = -.76$$

2. (a) Mean, (b) Median and Mode.

3. This means that 80% of the total group received language achievement scores below John Jones, and 20% received scores higher than his.

4. 25th percentile, 50th percentile, 75th percentile, Q_1, Q_2, Q_3.

5. 10th percentile, 40th percentile, 60th percentile, D_1, D_4, D_6.

6. (a) 40th percentile $= 11.5 + \left(\dfrac{.4(20) - 7}{6} \right) 1 = 11.67$

   $Q_3 = $ 75th percentile $= 12.5 + \left(\dfrac{.75(20) - 13}{4} \right) 1 = 13.00$

   $D_6 = $ 60th percentile $= 11.5 + \left(\dfrac{.60(20) - 7}{6} \right) 1 = 12.33$

   (b) 40th percentile $= 8.5 + \left(\dfrac{.40(25) - 6}{9} \right) 4 = 10.28$

   $Q_3 = $ 75th percentile $= 12.5 + \left(\dfrac{.75(25) - 15}{6} \right) 4 = 15$

   $D_6 = $ 60th percentile $= 8.5 + \left(\dfrac{.60(25) - 6}{9} \right) 4 = 12.5$

7. For score value 11: $\dfrac{\left(\dfrac{11 - 10.5}{1} \right) 3 + 4}{20} = $ 27.5th percentile.

   For score value 13: $\dfrac{\left(\dfrac{13 - 12.5}{1} \right) 4 + 13}{20} = $ 75th percentile.

For score value 14: $\dfrac{\left(\dfrac{14 - 13.5}{1}\right) 2 + 17}{20}$ = 90th percentile.

8. For score value 6: $\dfrac{\left(\dfrac{6 - 4.5}{4}\right) 4 + 2}{25}$ = 14th percentile

For score value 12: $\dfrac{\left(\dfrac{12 - 8.5}{4}\right) 9 + 6}{25}$ = 55.5th percentile.

For score value 17: $\dfrac{\left(\dfrac{17 - 16.5}{4}\right) 4 + 21}{25}$ = 86th percentile.

Set 5

1.

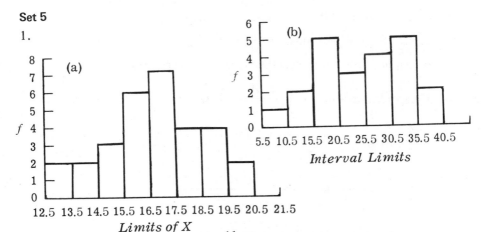

2.

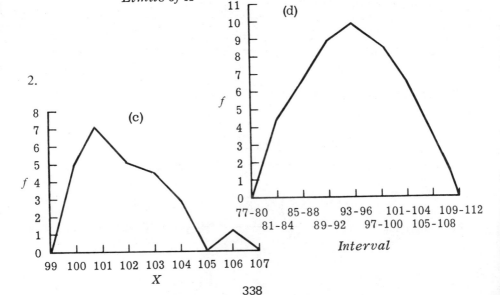

3. 1 (a), 2 (a), 2 (b)

4. 1 (b)

5. 2 (a) This frequency polygon is skewed to the right because the slope of the curve trails off in that direction.

6. 2 (b)

7. The mean is most affected by the degree of skewness in a distribution. The mode is not affected.

Set 6

1. (a) Range = 7, Q = 1.27; (b) Range = 34, Q = 7.00; (c) Range = 6, Q = 1.13; (d) Range = 27, Q = 5.11.

2. (a) Yes. The range will be decreased by the same amount as the extreme score is decreased. (b) No. The semi-interquartile range will remain the same. The value of an extreme score does not affect it.

3. The symbol x represents a deviation score. The symbol $|x|$ represents an absolute deviation score.

4. The term *absolute deviation score* means the amount that a raw score deviates from the $\bar{X}$, disregarding the direction of the deviation. It is calculated by subtracting the mean from the value of the raw score and disregarding the sign.

5. (a)

X	f	fX	x	$|x|$	$f|x|$
10	1	10	3	3	3
9	2	18	2	2	4
8	5	40	1	1	5
7	4	28	0	0	0
6	3	18	−1	1	3
5	3	15	−2	2	6
4	1	4	−3	3	3
	$N = 19$	133			$\Sigma f|x| = 24$

$\bar{X} = 7$

$$\text{A.D.} = \frac{24}{19} = 1.26$$

(b)

X	f	fX	x	$|x|$	$f|x|$
51	1	51	3.5	3.5	3.5
50	1	50	2.5	2.5	2.5
49	1	49	1.5	1.5	1.5
48	4	192	0.5	0.5	2.0
47	3	141	−0.5	0.5	1.5
46	2	92	−1.5	1.5	3.0
45	2	90	−2.5	2.5	5.0
	$N = 14$	665			$\Sigma f|x| = 19.0$

$\bar{X} = 47.5$

$$\text{A.D.} = \frac{19}{14} = 1.36$$

1.

X	f	fX	x	x^2	fx^2	
10	1	10	4	16	16	
9	2	18	3	9	18	
8	3	24	2	4	12	
7	4	28	1	1	4	
6	5	30	0	0	0	$\bar{X} = 6$
5	4	20	−1	1	4	
4	3	12	−2	4	12	$s'^2 = \dfrac{100}{25} = 4$
3	2	6	−3	9	18	
2	1	2	−4	16	16	$s' = \sqrt{4} = 2$
	$N = 25$	150			$\Sigma fx^2 = 100$	

2.

X	f	fX	X^2	fX^2	
20	1	20	400	400	
19	4	76	361	1444	$\bar{X} = 17.76$
18	6	108	324	1944	
17	3	51	289	867	$s'^2 = \dfrac{5392}{17} - (17.76)^2 = 1.76$
16	2	32	256	512	
15	1	15	225	225	$s' = \sqrt{1.76} = 1.33$
	$N = 17$	302		$\Sigma fX^2 = 5392$	

3. The "relative deviate" denotes the amount a raw score deviates from the mean expressed in standard deviation units. The symbol for the "relative deviate" is z.

4. For score 9 in 1, $z = 1.5$; for score 16 in 2, $z = -1.32$.

Set 8

1. The term *population* is used to denote all of the individuals with a given characteristic. The term *sample* is used to denote any portion of a population.

2. Population characteristics are called *parameters*. Sample characteristics are called *statistics*.

3. Population parameters are represented by Greek letters. Sample statistics are represented by Roman letters.

4. In a normal distribution, the μ, *Mdn*, and Mode are identical and are located at the modal point of the distribution.

5. (a) 34.13%, (b) 68.26%, (c) 4.54%, (d) 0.11%.

6. The symbol μ represents the mean of a population. The symbol σ^2 represents the variance of a population. The symbol σ represents the standard deviation of a population.

Set 9

1. The symbol P represents probability.

2. $P = .5$

3. $P = .15$

4. $P = .05$

5. (a) $P = .3413$, (b) $P = .4834$, (c) $P = .7187$, (d) $P = .0279$

Set 10

1. The "standard error of the mean" indicates the amount of variability among sample means expressed in standard deviation units. Its symbol is $\sigma_{\bar{x}}$.

2. (a) $P = .4772$, (b) $P = .8185$, (c) $P = .6713$

3. (a) $P = .3413$, (b) $P = .4082$, (c) $P = .3779$

4. The standard error of the mean is reduced when the size of the sample is increased.

Set 11

1. (a) $\Sigma x^2 = 66$. Identical answers for deviation score method and raw score method. (b) $\Sigma x^2 = 1170$. Identical answers for deviation score method and raw score method.

2. (a) $df = 15$; (b) $df = 19$.

3. (a) $s^2 = \dfrac{66}{15} = 4.40$; (b) $s^2 = \dfrac{1170}{19} = 61.58$

4. (a) $s^2 = \dfrac{16(850) - (112)^2}{16(15)} = 4.40$; (b) $s^2 = \dfrac{20(5670) - (300)^2}{20(19)} = 61.58$

5. s^2 = estimate of a population variance. σ^2 = population variance. df = degrees of freedom.

Set 12

1. (a) $s_{\bar{X}} = .5$; (b) $s_{\bar{X}} = 1.75$

2. 95% confidence interval is from 14.02 to 15.98.

3. 95% confidence interval is from 196.57 to 203.43.

4. Researchers generally will accept a hypothesis as being correct if the probability of its being incorrect is only $P = .05$.

5. Estimate of the standard error of the mean.

Set 13

1. The 95% and the 99% confidence intervals. The 99% confidence interval is the more stringent.

2. (a) 99% confidence interval, 98.27 to 101.73; (b) 99% confidence interval, 73.32 to 76.68.

3. (a) 95% confidence interval, 42.67 to 57.33. 99% confidence interval, 39.99 to 60.01. (b) 95% confidence interval, 23.55 to 41.65. 99% confidence interval, 19.83 to 45.37.

4. The t distribution provides probabilities associated with differences between μ and $\bar{X}$ where $s_{\bar{X}}$ has been estimated from the data in a small sample. The shape of the t distribution varies with the degrees of freedom associated with the sample. As the size of the sample becomes larger, the t distribution more nearly resembles the normal distribution.

Set 14

1. There is no difference in the final examination scores of freshmen who are taught elementary statistics by the two different methods.

2. There is no difference between the final algebra examination scores of children who study in the morning and those who study in the evening.

3. The standard error of the difference is the standard deviation of differences between sample means.

4. You should accept the null hypothesis.

5. Accepting the null hypothesis means that the difference between your two sample means is a result of sampling error and does not represent a real difference.

6. The symbol $\sigma_{\bar{X}_1 - \bar{X}_2}$ represents the standard error of the difference between sample means.

Set 15

1. $s_{\bar{X}_1 - \bar{X}_2} = .94$, $t = -2.4/.94 = -2.55$

2. For $df = 30$, $P = .05$, $t = 2.042$; $P = .01$, $t = 2.750$. Because the t ratio exceeds that required for $P = .05$, the null hypothesis should be rejected at that level. Because the t ratio does not exceed that required for $P = .01$, the null hypothesis should not be rejected at that level.

3. $s_{\bar{X}_1 - \bar{X}_2} = 1.00$, $t = -3.3/1.00 = -3.3$

4. For $df = 26$, $P = .05$, $t = 2.056$; $P = .01$, $t = 2.779$. Because the t ratio exceeds that required for both $P = .05$ and $P = .01$, the null hypothesis should be rejected at $P = .01$.

5. $s_{\bar{X}_1 - \bar{X}_2}$ is the symbol for the estimate of the standard error of the difference between sample means.

Set 16

1. Scores are considered non-independent when they are for the same group of individuals or when the individuals have been matched on some variable.

2. $s_D^2 = 7.73$, $s_{\bar{D}} = .879$, $t = .227$

3. At the .05 level of significance, the null hypothesis should be accepted. This test requires $t = 2.262$ to be significant at $P = .05$.

4. $s_{\bar{D}} = 1.00$, $t = 6.00$

5. At $df = 7$ for significance at $P = .05$, the t ratio must be at least 2.365. Therefore, the null hypothesis should be rejected.

6. The symbol D represents the difference between scores on the same variable for two individuals who have been matched on some other variable or the difference between two scores for the same individual.

Set 17

1. A one-tailed test is appropriate when the direction of the difference is hypothesized. A two-tailed test is appropriate when the direction of the difference is not hypothesized.

2. (a) One-tailed test. (b) One-tailed test. (c) Two-tailed test. (d) Two-tailed test.

3. For a two-tailed test, $t = 2.179$; for a one-tailed test, $t = 1.782$.

4. For a two-tailed test, $t = 2.807$; for a one-tailed test, $t = 2.500$.

5. A two-tailed test requires a larger t ratio than a one-tailed test.

6. A Type I error occurs if you reject the null hypothesis when, in fact, no difference exists.

7. A Type II error occurs if you accept the null hypothesis when, in fact, a true difference exists.

8. The smaller the level of significance the less probability there is that you have made a Type I error if you reject the null hypothesis. The smaller the level of significance, the more probability there is that you have made a Type II error if you accept the null hypothesis.

Set 18

1. The total sum of squares is divided into the sum of squares within groups and the sum of squares between groups.

2. $SS_t = 611; SS_b = 180; SS_w = 431$.

3. Between groups: $df_b = 2, MS_b = 90$. Within groups: $df_w = 27, MS_w = 15.96$.

4. The symbol SS_b represents the sum of squares between groups; the symbol SS_w represents the sum of squares within groups; the symbol SS_t represents the total sum of squares; the symbol MS represents the Mean Square, which is the term used to describe the variance estimate in the analysis-of-variance technique.

Set 19

1. $F = 90/15.96 = 5.64$

2. Entering Table 3 with 2 and 25 degrees of freedom, for significance at $P = .01$ an F ratio of 5.57 is required. The obtained F ratio of 5.64 exceeds the required value for significance at $P = .01$. On the basis of the F test, you should reject the null hypothesis at $P = .01$.

3. $F = 215/72 = 2.986$. For significance at $P = .02$, at df's of 24 and 40, an F ratio of 2.29 is required. Therefore, the null hypothesis that there is no difference in the variability of the scores in the two groups should be rejected.

4. The symbol F represents the ratio between two variance estimates.

Set 20

1.

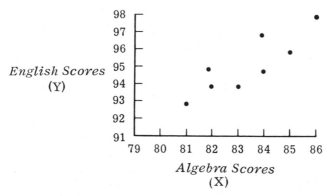

2. A perfect positive correlation between two variables means that for every increase in score value on one variable there is a proportionate increase in score value on the other variable. A moderate positive correlation between two variables means that every increase in score value on one variable is not perfectly matched with a proportionate increase on the other variable, although they generally increase.

3. A perfect negative correlation between two variables means that for every increase in score value on one variable there is a corresponding decrease in score value on the other variable. A moderate negative correlation between two variables means that every increase in score value on one variable is not perfectly matched with a corresponding decrease on the other variable, although they generally decrease.

4. A positive correlation. It is not perfect.

5. A zero correlation means that the scores on one variable are not related, either negatively or positively, to the scores on the other variable.

Set 21

1. Perfect positive correlation: $r = 1.00$; perfect negative correlation: $r = -1.00$

2. $r = \dfrac{8(1846) - (120)(116)}{\sqrt{[8(1886) - (120)^2]\,[8(1916) - (116)^2]}} = .747$

3. It requires a two-tailed test of significance. For $df = 6$; $P = .05$, $r = .707$, $P = .01$, $r = .834$

4. There is no relationship between arithmetic scores and English scores. Because the obtained r does exceed that required at $P = .05$, the null hypothesis should be rejected at that level.

5. $r = \dfrac{11(3757) - (276)(142)}{\sqrt{[11(7152) - (276)^2]\,[11(2080) - (142)^2]}} = .82$

345

6. It requires a one-tailed test of significance. For $df = 9$: $P = .05$, $r = .521$; $P = .01$, $r = .685$

7. There is no relationship between spelling scores and English scores. Because the obtained r exceeds that required at $P = .01$, the null hypothesis should be rejected at that level.

8. The symbol r represents the Pearson product-moment correlation coefficient.

Set 22

1. A regression line is the line drawn in a scatter diagram which "best fits" the correlated data for two variables. It is the line from which you can predict the score value on one variable for any selected value on the other variable.

2. Regression of Y on X: $b_{yx} = \dfrac{833 - \dfrac{(150)(52)}{10}}{2320 - \dfrac{(150)^2}{10}} = .757$

3. For regression of Y on X: $a = 5.2 - (.757)15 = -6.155$. For score 13: $\tilde{Y} = -6.155 + (.757)(13) = 3.686$. For score 19: $\tilde{Y} = -6.155 + (.757)(19) = 8.228$.

4.

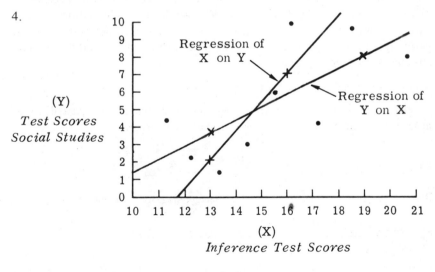

(Y)

Test Scores
Social Studies

(X)

Inference Test Scores

5. Regression of X on Y: $b_{xy} = \dfrac{833 - \dfrac{(150)(52)}{10}}{352 - \dfrac{(52)^2}{10}} = .650$

6. For regression of X on Y: $a = 15 - (.650)5.2 = 11.620$. For score 2: $\widetilde{X} = 11.620 + (.650)(2) = 12.920$. For score 7: $\widetilde{X} = 11.620 + (.650)$ $(7) = 16.170$.

7. See scatter diagram in Exercise 2.

8. The symbol $\widetilde{Y}$ represents a value of Y predicted by the regression equation for Y on X. The symbol b_{yx} represents the regression coefficient for Y on X. The symbol $\widetilde{X}$ represents a value of X predicted by the regression equation for X on Y. The symbol b_{xy} represents the regression coefficient for X on Y.

Set 23

1. $\rho = 1 - \dfrac{6(40)}{12(144 - 1)} = .860$

2. $N = 12$. For $P = .01$, $\rho = .780$. The rho computed in Exercise 1 exceeds the value required for significance at $P = .01$. Therefore, the null hypothesis that the number of hours of marksmanship practice are not related to proficiency ratings should be rejected.

3. $\rho = 1 - \dfrac{6(202)}{10(100 - 1)} = -.224$

4. $N = 10$. For $P = .05$, $\rho = .648$. The rho computed in Exercise 3 does not reach the value required for significance at $P = .05$. Therefore, you should accept the null hypothesis that the teacher ratings of the children's cooperativeness and the children's self-ratings are not related.

5. The symbol rho (ρ) represents Spearman's rank order correlation.

Set 24

1. $\chi^2 = \dfrac{(8 - .5)^2}{27} + \dfrac{(8 - .5)^2}{27} = 4.16$

2. $df = 1$. For $P = .05$, $\chi^2 = 3.84$. The obtained chi square in Exercise 1 exceeds the value required at $P = .05$. Therefore, the null hypothesis should be rejected.

3. $\chi^2 = \dfrac{(12.5)^2}{22.5} + \dfrac{(9.5)^2}{22.5} + \dfrac{(4.5)^2}{22.5} + \dfrac{(17.5)^2}{22.5} = 25.47$

4. $df = 3$. For $P = .05$, $\chi^2 = 7.82$. The obtained chi square in Exercise 3 does exceed the value required at $P = .05$. Therefore, the null hypothesis should be rejected.

5. $\chi^2 = \dfrac{(7.7)^2}{13.3} + \dfrac{(2.3)^2}{13.3} + \dfrac{(5.3)^2}{13.3} = 6.966$

6. $df = 2$. For $P = .05$, $\chi^2 = 5.99$. The obtained chi square in Exercise 5 exceeds the value required at $P = .05$. Therefore, the null hypothesis should be rejected. You conclude that the installation of an employee's lounge is associated with increased productivity.

7. $\chi^2 = \dfrac{(20.5)^2}{150} + \dfrac{(20.5)^2}{150} = 5.60$

8. $df = 1$. For $P = .01$, $\chi^2 = 6.64$. The obtained chi square in Exercise 7 does not exceed the value required at $P = .01$. Therefore, the null hypothesis should be accepted.

9. The symbol χ^2 represents chi square.

Set 25

1. $\chi^2 = 9.58$

2. $df = 2$. For $P = .05$, $\chi^2 = 5.99$. The obtained chi square in Exercise 1 exceeds the value required at $P = .05$. Therefore, the null hypothesis should be rejected. The data indicates that boys have a preference for basketball and do not prefer volleyball, whereas girls have a preference for volleyball and do not prefer basketball. There appears to be little difference in their preference of kickball.

3. $\chi^2 = 4.037$

4. $df = 12$. For $P = .05$, $\chi^2 = 21.03$. The obtained chi square in Exercise 3 does not exceed the value required at $P = .05$. Therefore, the null hypothesis should be accepted. Your conclusion should be that there is no difference between types of television programs preferred by different nationality groups.

FORMULAS

Formula 1. Calculation of the Median

$$Mdn = \ell\ell + \left(\frac{.5N - \Sigma f_b}{f_w} \right) i$$

in which Σ = the sum of
        f_b = frequency below the interval which contains the Mdn
        f_w = frequency within the interval which contains the Mdn

Formula 2. Calculation of the Mean

$$\bar{X} = \frac{\Sigma fX}{N}$$

Formula 3. Calculation of the Percentile

Percentile in decimal form $\dfrac{\left(\dfrac{X - \ell\ell}{i} \right) f_w + \Sigma f_b}{N}$

in which X = score value for which the percentile is to be computed
              f_b = frequency below the i which contains the score
              f_w = frequency within the i which contains the score

Multiply the decimal form of the percentile by 100 in order to determine the percentile.

Formula 4. Calculation of the Range

$$\text{Range} = H - L$$

in which H = highest score in frequency distribution
              L = lowest score in frequency distribution

Formula 5. Calculation of the Semi-interquartile Range

$$Q = \frac{Q_3 - Q_1}{2}$$

Formula 6. Calculation of a Deviation Score

$$x = X - \bar{X}$$

Formula 7. Calculation of the Average Deviation

$$\text{A.D.} = \frac{\Sigma f|x|}{N}$$

in which $|x|$ = the absolute deviation of a raw score from the mean

---

Formula 8. Calculation of the Sample Variance

$$\text{Deviation Score Method} \quad s'^2 = \frac{\Sigma f x^2}{N} \qquad \text{(Formula 8a)}$$

$$\text{Raw Score Method} \quad s'^2 = \frac{\Sigma f X^2}{N} - \bar{X}^2 \qquad \text{(Formula 8b)}$$

Formula 9. Calculation of the Sample Standard Deviation

$$s' = \sqrt{s'^2}$$

Formula 10. Calculation of a Relative Deviate

$$z = \frac{x}{s'}$$

Formula 11. Calculation of the Sum of Squared Deviations ("Sum of Squares") (Formulas 11a and 11b are Equivalent)[*]

Deviation Score Method $\Sigma x^2 = \Sigma(X - \bar{X})^2$ (Formula 11a)

Raw Score Method $\Sigma x^2 = \Sigma X^2 - \dfrac{(\Sigma X)^2}{N}$ (Formula 11b)

Formula 12. Estimate of the Population Variance from Sample Data (Formulas 12a and 12b are Equivalent)

Deviation Score Method $s^2 = \dfrac{\Sigma x^2}{N - 1}$ (Formula 12a)

Raw Score Method $s^2 = \dfrac{N\Sigma X^2 - (\Sigma X)^2}{N(N - 1)}$ (Formula 12b)

Formula 13. Estimate of the Standard Error of the Mean

$$s_{\bar{X}} = \frac{s}{\sqrt{N}}$$

---

[*]The symbol f is omitted in these and subsequent formulas. It is to be assumed that all frequencies in the distribution are used in the calculations.

Formula 14. Estimation of the Common Population Variance (Pooled Variance) from the Data in Two Samples (Formulas 14a and 14b are Equivalent)

$$s^2 = \frac{\Sigma x_1^2 + \Sigma x_2^2}{N_1 + N_2 - 2}$$

(Formula 14a)

$$s^2 = \frac{(N_1 - 1)s_1^2 + (N_2 - 1)s_2^2}{N_1 + N_2 - 2}$$

(Formula 14b)

Formula 15. Estimate of the Standard Error of the Difference between Means (Pooled Variance Method)

$$s_{\bar{X}_1 - \bar{X}_2} = \sqrt{\frac{s^2}{N_1} + \frac{s^2}{N_2}}$$

Formula 16. Calculation of the t Ratio for Independent Means

$$t = \frac{\bar{X}_1 - \bar{X}_2}{s_{\bar{X}_1 - \bar{X}_2}}$$

Degrees of Freedom: $N_1 + N_2 - 2$

Formula 17. Estimation of the Population Variance of Difference Scores

$$s_D^2 = \frac{N\Sigma D^2 - (\Sigma D)^2}{N(N - 1)}$$

in which $D = X_1 - X_2$ for each pair of scores

Formula 18. Estimation of the Population Standard Error of the Mean Difference Scores (Formulas 18a and 18b are Equivalent)

$$s_{\bar{D}} = \sqrt{\frac{s_D^2}{N}} \qquad \text{(Formula 18a)}$$

$$s_{\bar{D}} = \sqrt{\frac{N\Sigma D^2 - (\Sigma D)^2}{N^2(N-1)}} \qquad \text{(Formula 18b)}$$

in which N = number of pairs of scores

Formula 19. Calculation of the t ratio for Non-independent Means

$$t = \frac{\bar{X}_1 - \bar{X}_2}{s_{\bar{D}}}$$

Degrees of freedom: $N - 1$ pairs of scores

Formula 20. F Test. Formula for Computing the F Ratio in the Analysis of Variance

$$F = \frac{\text{Mean Square between groups}}{\text{Mean Square within groups}} = \frac{MS_b}{MS_w}$$

Formula 21. Composition of the Total Sum of Squares

Total sum of squares = sum of squares between groups + sum of squares within groups.

$$SS_t = SS_b + SS_w$$

Formulas 22 through 24. Calculation of Sum of Squares

Total Sum of Squares

$$SS_t = \Sigma X^2 - \frac{(\Sigma X)^2}{N_t} \qquad \text{(Formula 22)}$$

Sum of Squares within Groups

$$SS_w = SS_A + SS_B + SS_C \qquad \text{(Formula 23)}$$
$$\text{(within group)}$$

Sum of Squares between Groups

$$SS_b = SS_t - SS_w \qquad \text{(Formula 24a)}$$

$$SS_b = N_A(\bar{X}_A - \bar{X}_t)^2 + N_B(\bar{X}_B - \bar{X}_t)^2 + N_C(\bar{X}_C - \bar{X}_t)^2 \qquad \text{(Formula 24b)}$$

Formulas 25 through 27. Calculation of Degrees of Freedom for Analysis of Variance

Degrees of freedom

Total

$$df_t = N - 1 \qquad \text{(Formula 25)}$$

Between groups

$$df_b = \text{No. of groups} - 1 \qquad \text{(Formula 26)}$$

Within groups

$$df_w = df_t - df_b \qquad \text{(Formula 27)}$$

Formulas 28 and 29. Calculation of Mean Squares (variance estimates)

$$MS_b = \frac{SS_b}{df_b} \qquad \text{(Formula 28)}$$

$$MS_w = \frac{SS_w}{df_w} \qquad \text{(Formula 29)}$$

Formula 30. Computational Formulas for Sums of Squares

Group A	Group B	Group C	Total
ΣX_A	ΣX_B	ΣX_C	ΣX_t
$\Sigma X_A{}^2$	$\Sigma X_B{}^2$	$\Sigma X_C{}^2$	$\Sigma X_t{}^2$
N_A	N_B	N_C	N_t

Step 1   Correction term

$$C = \frac{(\Sigma X_t)^2}{N_t}$$

Step 2   Total sum of squares

$$SS_t = \Sigma X_t{}^2 - C$$

Step 3   Sum of squares between groups

$$SS_b = \frac{(\Sigma X_A)^2}{N_A} + \frac{(\Sigma X_B)^2}{N_B} + \frac{(\Sigma X_C)^2}{N_C} - C$$

Step 4   Sum of squares within groups

$$SS_w = SS_t - SS_b$$

Formula 31. F Test. Calculation of the F Ratio for Comparison of Two Variance Estimates

$$F = \frac{\text{larger } s^2}{\text{smaller } s^2}$$

Formula 32. Calculation of the Pearson Product-Moment Correlation Coefficient

$$r = \frac{N\Sigma XY - (\Sigma X)(\Sigma Y)}{\sqrt{[N\Sigma X^2 - (\Sigma X)^2]\,[N\Sigma Y^2 - (\Sigma Y)^2]}}$$

in which N = number of pairs of scores

Degrees of Freedom: $N - 2$

Formulas 33-35. Formulas for Calculation of the Regression of Y on X

Formula 33. $\qquad\qquad \tilde{Y} = a + b_{yx}X \qquad\qquad$ (regression equation)

Formula 34. $\qquad b_{yx} = \dfrac{\Sigma XY - \dfrac{(\Sigma X)(\Sigma Y)}{N}}{\Sigma X^2 - \dfrac{(\Sigma X)^2}{N}} \qquad$ (regression coefficient)

Formula 35. $\qquad\qquad a = \bar{Y} - b_{yx}\bar{X}$

in which $\tilde{Y}$ is the predicted value of Y

Formulas 36-38. Formulas for Calculation of the Regression of X on Y

Formula 36. $\qquad\qquad \tilde{X} = a + b_{xy}Y \qquad\qquad$ (regression equation)

Formula 37. $\qquad b_{xy} = \dfrac{\Sigma XY - \dfrac{(\Sigma X)(\Sigma Y)}{N}}{\Sigma Y^2 - \dfrac{(\Sigma Y)^2}{N}} \qquad$ (regression coefficient)

Formula 38. $\qquad\qquad a = \bar{X} - b_{xy}\bar{Y}$

where $\tilde{X}$ is the predicted value of X

Formula 39. Calculation of Spearman's Rank Order
Correlation Coefficient (rho)

$$\rho = 1 - \frac{6\Sigma D^2}{N(N^2 - 1)}$$

in which $\quad D$ = difference between a pair of ranks

$\qquad\qquad N$ = number of pairs of ranks

Formula 40. Calculation of the Chi Square when *df* = 1

$$\chi^2 = \sum \frac{(|O - E| - .5)^2}{E}$$

in which O = observed frequency

         E = expected frequency

The subtraction of .5 from each $|O - E|$ represents the Yates' correction for continuity

---

Formula 41. Calculation of the Chi Square when *df* is Larger than 1

$$\chi^2 = \sum \frac{(O - E)^2}{E}$$

Formula 42. Calculation of the Expected Frequency (*E*) of a Cell

$$E = \frac{(N_{row})(N_{col})}{N_{total}}$$

Degrees of Freedom: (number of rows − 1) (number of columns − 1)

TABLES

TABLE 1. AREAS OF THE STANDARD NORMAL CURVE

z	μ to z	z	μ to z	z	μ to z	z	μ to z	z	μ to z	z	μ to z	z	μ to z	z	μ to z
0.00	.0000	0.25	.0987	0.50	.1915	0.75	.2734	1.00	.3413	1.25	.3944	1.50	.4332	1.75	.4599
0.01	.0040	0.26	.1026	0.51	.1950	0.76	.2764	1.01	.3438	1.26	.3962	1.51	.4345	1.76	.4608
0.02	.0080	0.27	.1064	0.52	.1985	0.77	.2794	1.02	.3461	1.27	.3980	1.52	.4357	1.77	.4616
0.03	.0120	0.28	.1103	0.53	.2019	0.78	.2823	1.03	.3485	1.28	.3997	1.53	.4370	1.78	.4625
0.04	.0160	0.29	.1141	0.54	.2054	0.79	.2852	1.04	.3508	1.29	.4015	1.54	.4382	1.79	.4633
0.05	.0199	0.30	.1179	0.55	.2088	0.80	.2881	1.05	.3531	1.30	.4032	1.55	.4394	1.80	.4641
0.06	.0239	0.31	.1217	0.56	.2123	0.81	.2910	1.06	.3554	1.31	.4049	1.56	.4406		
0.07	.0279	0.32	.1255	0.57	.2157	0.82	.2939	1.07	.3577	1.32	.4066	1.57	.4418		
0.08	.0319	0.33	.1293	0.58	.2190	0.83	.2967	1.08	.3599	1.33	.4082	1.58	.4429		
0.09	.0359	0.34	.1331	0.59	.2221	0.84	.2995	1.09	.3621	1.34	.4099	1.59	.4441		
0.10	.0398	0.35	.1368	0.60	.2257	0.85	.3023	1.10	.3643	1.35	.4115	1.60	.4452		
0.11	.0438	0.36	.1406	0.61	.2291	0.86	.3051	1.11	.3665	1.36	.4131	1.61	.4463		
0.12	.0478	0.37	.1443	0.62	.2324	0.87	.3078	1.12	.3686	1.37	.4147	1.62	.4474		
0.13	.0517	0.38	.1480	0.63	.2357	0.88	.3106	1.13	.3708	1.38	.4162	1.63	.4484		
0.14	.0557	0.39	.1517	0.64	.2389	0.89	.3133	1.14	.3729	1.39	.4177	1.64	.4495		
0.15	.0596	0.40	.1554	0.65	.2422	0.90	.3159	1.15	.3749	1.40	.4192	1.65	.4505		
0.16	.0636	0.41	.1591	0.66	.2454	0.91	.3186	1.16	.3770	1.41	.4207	1.66	.4515		
0.17	.0675	0.42	.1628	0.67	.2486	0.92	.3212	1.17	.3790	1.42	.4222	1.67	.4525		
0.18	.0714	0.43	.1664	0.68	.2517	0.93	.3238	1.18	.3810	1.43	.4236	1.68	.4535		
0.19	.0753	0.44	.1700	0.69	.2549	0.94	.3264	1.19	.3830	1.44	.4251	1.69	.4545		
0.20	.0793	0.45	.1736	0.70	.2580	0.95	.3289	1.20	.3849	1.45	.4265	1.70	.4554		
0.21	.0832	0.46	.1772	0.71	.2611	0.96	.3315	1.21	.3869	1.46	.4279	1.71	.4564		
0.22	.0871	0.47	.1808	0.72	.2642	0.97	.3340	1.22	.3888	1.47	.4292	1.72	.4573		
0.23	.0910	0.48	.1844	0.73	.2673	0.98	.3365	1.23	.3907	1.48	.4306	1.73	.4582		
0.24	.0948	0.49	.1879	0.74	.2704	0.99	.3389	1.24	.3925	1.49	.4319	1.74	.4591		
0.25	.0987	0.50	.1915	0.75	.2734	1.00	.3413	1.25	.3944	1.50	.4332	1.75	.4599		

z	μ to z	z	μ to z	z	μ to z	z	μ to z	z	μ to z	z	μ to z
1.80	.4641	2.05	.4798	2.30	.4893	2.55	.4946	2.80	.4974	3.05	.4989
1.81	.4649	2.06	.4803	2.31	.4896	2.56	.4948	2.81	.4975	3.06	.4989
1.82	.4656	2.07	.4808	2.32	.4898	2.57	.4949	2.82	.4976	3.07	.4989
1.83	.4664	2.08	.4812	2.33	.4901	2.58	.4951	2.83	.4977	3.08	.4990
1.84	.4671	2.09	.4817	2.34	.4904	2.59	.4952	2.84	.4977	3.09	.4990
1.85	.4678	2.10	.4821	2.35	.4906	2.60	.4953	2.85	.4978	3.10	.4990
1.86	.4686	2.11	.4826	2.36	.4909	2.61	.4955	2.86	.4979	3.11	.4991
1.87	.4693	2.12	.4830	2.37	.4911	2.62	.4956	2.87	.4979	3.12	.4991
1.88	.4699	2.13	.4834	2.38	.4913	2.63	.4957	2.88	.4980	3.13	.4991
1.89	.4706	2.14	.4838	2.39	.4916	2.64	.4959	2.89	.4981	3.14	.4992
1.90	.4713	2.15	.4842	2.40	.4918	2.65	.4960	2.90	.4981	3.15	.4992
1.91	.4719	2.16	.4846	2.41	.4920	2.66	.4961	2.91	.4982	3.16	.4992
1.92	.4726	2.17	.4850	2.42	.4922	2.67	.4962	2.92	.4982	3.17	.4992
1.93	.4732	2.18	.4854	2.43	.4925	2.68	.4963	2.93	.4983	3.18	.4993
1.94	.4738	2.19	.4857	2.44	.4927	2.69	.4964	2.94	.4984	3.19	.4993
1.95	.4744	2.20	.4861	2.45	.4929	2.70	.4965	2.95	.4984	3.20	.4993
1.96	.4750	2.21	.4864	2.46	.4931	2.71	.4966	2.96	.4985	3.21	.4993
1.97	.4756	2.22	.4868	2.47	.4932	2.72	.4967	2.97	.4985	3.22	.4994
1.98	.4761	2.23	.4871	2.48	.4934	2.73	.4968	2.98	.4986	3.23	.4994
1.99	.4767	2.24	.4875	2.49	.4936	2.74	.4969	2.99	.4986	3.24	.4994
2.00	.4772	2.25	.4878	2.50	.4938	2.75	.4970	3.00	.4987	3.30	.4995
2.01	.4778	2.26	.4881	2.51	.4940	2.76	.4971	3.01	.4987	3.40	.4997
2.02	.4783	2.27	.4884	2.52	.4941	2.77	.4972	3.02	.4987	3.50	.4998
2.03	.4788	2.28	.4887	2.53	.4943	2.78	.4973	3.03	.4988	3.60	.4998
2.04	.4793	2.29	.4890	2.54	.4945	2.79	.4974	3.04	.4988	3.70	.4999
2.05	.4798	2.30	.4893	2.55	.4946	2.80	.4974	3.05	.4989		

Table 1 is abridged and reprinted from Tables III and VI, Allen L. Edwards, *Statistical Methods for the Behavioral Sciences*, Holt, Rinehart & Winston, Inc., New York, 1954. By permission of the author and publisher.

TABLE 2. DISTRIBUTION OF *t*

df	P = .10	P = .05	P = .02	P = .01
1	6.314	12.706	31.821	63.657
2	2.920	4.303	6.965	9.925
3	2.353	3.182	4.541	5.841
4	2.132	2.776	3.747	4.604
5	2.015	2.571	3.365	4.032
6	1.943	2.447	3.143	3.707
7	1.895	2.365	2.998	3.499
8	1.860	2.306	2.896	3.355
9	1.833	2.262	2.821	3.250
10	1.812	2.228	2.764	3.169
11	1.796	2.201	2.718	3.106
12	1.782	2.179	2.681	3.055
13	1.771	2.160	2.650	3.012
14	1.761	2.145	2.624	2.977
15	1.753	2.131	2.602	2.947
16	1.746	2.120	2.583	2.921
17	1.740	2.110	2.567	2.898
18	1.734	2.101	2.552	2.878
19	1.729	2.093	2.539	2.861
20	1.725	2.086	2.528	2.845
21	1.721	2.080	2.518	2.831
22	1.717	2.074	2.508	2.819
23	1.714	2.069	2.500	2.807
24	1.711	2.064	2.492	2.797
25	1.708	2.060	2.485	2.787
26	1.706	2.056	2.479	2.779
27	1.703	2.052	2.473	2.771
28	1.701	2.048	2.467	2.763
29	1.699	2.045	2.462	2.756
30	1.697	2.042	2.457	2.750
60	1.671	2.000	2.390	2.660
∞	1.645	1.960	2.326	2.576

Table 2 is taken from Table 3 of Fisher & Yates: *Statistical Tables for Biological, Agricultural and Medical Research,* published by Oliver & Boyd Limited, Edinburgh and by permission of the authors and publishers.

TABLE 3. TABLE OF F FOR .05 (ROMAN) and .01 (BOLDFACE) LEVELS OF SIGNIFICANCE

Degrees of Freedom for Greater Mean Square

Degrees of Freedom for Smaller Mean Square	1	2	3	4	5	6	8	12	24	∞
1	161.45 **4052.10**	199.50 **4999.03**	215.72 **5403.49**	224.57 **5625.14**	230.17 **5764.08**	233.97 **5859.39**	238.89 **5981.34**	243.91 **6105.83**	249.04 **6234.16**	254.32 **6366.48**
2	18.51 **98.49**	19.00 **99.01**	19.16 **99.17**	19.25 **99.25**	19.30 **99.30**	19.33 **99.33**	19.37 **99.36**	19.41 **99.42**	19.45 **99.46**	19.50 **99.50**
3	10.13 **34.12**	9.55 **30.81**	9.28 **29.46**	9.12 **28.71**	9.01 **28.24**	8.94 **27.91**	8.84 **27.49**	8.74 **27.05**	8.64 **26.60**	8.53 **26.12**
4	7.71 **21.20**	6.94 **18.00**	6.59 **16.69**	6.39 **15.98**	6.26 **15.52**	6.16 **15.21**	6.04 **14.80**	5.91 **14.37**	5.77 **13.93**	5.63 **13.46**
5	6.61 **16.26**	5.79 **13.27**	5.41 **12.06**	5.19 **11.39**	5.05 **10.97**	4.95 **10.67**	4.82 **10.27**	4.68 **9.89**	4.53 **9.47**	4.36 **9.02**
6	5.99 **13.74**	5.14 **10.92**	4.76 **9.78**	4.53 **9.15**	4.39 **8.75**	4.28 **8.47**	4.15 **8.10**	4.00 **7.72**	3.84 **7.31**	3.67 **6.88**
7	5.59 **12.25**	4.74 **9.55**	4.35 **8.45**	4.12 **7.85**	3.97 **7.46**	3.87 **7.19**	3.73 **6.84**	3.57 **6.47**	3.41 **6.07**	3.23 **5.65**
8	5.32 **11.26**	4.46 **8.65**	4.07 **7.59**	3.84 **7.01**	3.69 **6.63**	3.58 **6.37**	3.44 **6.03**	3.28 **5.67**	3.12 **5.28**	2.93 **4.86**
9	5.12 **10.56**	4.26 **8.02**	3.86 **6.99**	3.63 **6.42**	3.48 **6.06**	3.37 **5.80**	3.23 **5.47**	3.07 **5.11**	2.90 **4.73**	2.71 **4.31**
10	4.96 **10.04**	4.10 **7.56**	3.71 **6.55**	3.48 **5.99**	3.33 **5.64**	3.22 **5.39**	3.07 **5.06**	2.91 **4.71**	2.74 **4.33**	2.54 **3.91**
11	4.84 **9.65**	3.98 **7.20**	3.59 **6.22**	3.36 **5.67**	3.20 **5.32**	3.09 **5.07**	2.95 **4.74**	2.79 **4.40**	2.61 **4.02**	2.40 **3.60**
12	4.75 **9.33**	3.88 **6.93**	3.49 **5.95**	3.26 **5.41**	3.11 **5.06**	3.00 **4.82**	2.85 **4.50**	2.69 **4.16**	2.50 **3.78**	2.30 **3.36**
14	4.60 **8.86**	3.74 **6.51**	3.34 **5.56**	3.11 **5.03**	2.96 **4.69**	2.85 **4.46**	2.70 **4.14**	2.53 **3.80**	2.35 **3.43**	2.13 **3.00**
16	4.49 **8.53**	3.63 **6.23**	3.24 **5.29**	3.01 **4.77**	2.85 **4.44**	2.74 **4.20**	2.59 **3.89**	2.42 **3.55**	2.24 **3.18**	2.01 **2.75**

Degrees of Freedom for Smaller Mean Square	1	2	3	4	5	6	8	12	24	∞
18	4.41 / **8.28**	3.55 / **6.01**	3.16 / **5.09**	2.93 / **4.58**	2.77 / **4.25**	2.66 / **4.01**	2.51 / **3.71**	2.34 / **3.37**	2.15 / **3.01**	1.92 / **2.57**
20	4.35 / **8.10**	3.49 / **5.85**	3.10 / **4.94**	2.87 / **4.43**	2.71 / **4.10**	2.60 / **3.87**	2.45 / **3.56**	2.28 / **3.23**	2.08 / **2.86**	1.84 / **2.42**
25	4.24 / **7.77**	3.38 / **5.57**	2.99 / **4.68**	2.76 / **4.18**	2.60 / **3.86**	2.49 / **3.63**	2.34 / **3.32**	2.16 / **2.99**	1.96 / **2.62**	1.71 / **2.17**
30	4.17 / **7.56**	3.32 / **5.39**	2.92 / **4.51**	2.69 / **4.02**	2.53 / **3.70**	2.42 / **3.47**	2.27 / **3.17**	2.09 / **2.84**	1.89 / **2.47**	1.62 / **2.01**
40	4.08 / **7.31**	3.23 / **5.18**	2.84 / **4.31**	2.61 / **3.83**	2.45 / **3.51**	2.34 / **3.29**	2.18 / **2.99**	2.00 / **2.66**	1.79 / **2.29**	1.52 / **1.82**
50	4.03 / **7.17**	3.18 / **5.06**	2.79 / **4.20**	2.56 / **3.72**	2.40 / **3.41**	2.29 / **3.19**	2.13 / **2.89**	1.95 / **2.56**	1.74 / **2.18**	1.44 / **1.68**
60	4.00 / **7.08**	3.15 / **4.98**	2.76 / **4.13**	2.52 / **3.65**	2.37 / **3.34**	2.25 / **3.12**	2.10 / **2.82**	1.92 / **2.50**	1.70 / **2.12**	1.39 / **1.60**
70	3.98 / **7.01**	3.13 / **4.92**	2.74 / **4.07**	2.50 / **3.60**	2.35 / **3.29**	2.23 / **3.07**	2.07 / **2.78**	1.89 / **2.45**	1.67 / **2.07**	1.35 / **1.53**
80	3.96 / **6.96**	3.11 / **4.88**	2.72 / **4.04**	2.49 / **3.56**	2.33 / **3.26**	2.21 / **3.04**	2.06 / **2.74**	1.88 / **2.42**	1.65 / **2.03**	1.31 / **1.47**
90	3.95 / **6.92**	3.10 / **4.85**	2.71 / **4.01**	2.47 / **3.53**	2.32 / **3.23**	2.20 / **3.01**	2.04 / **2.72**	1.86 / **2.39**	1.64 / **2.00**	1.28 / **1.43**
100	3.94 / **6.90**	3.09 / **4.82**	2.70 / **3.98**	2.46 / **3.51**	2.30 / **3.21**	2.19 / **2.99**	2.03 / **2.69**	1.85 / **2.37**	1.63 / **1.98**	1.26 / **1.39**
200	3.89 / **6.97**	3.04 / **4.71**	2.65 / **3.88**	2.42 / **3.41**	2.26 / **3.11**	2.14 / **2.89**	1.98 / **2.60**	1.80 / **2.28**	1.57 / **1.88**	1.14 / **1.21**
∞	3.84 / **6.64**	2.99 / **4.60**	2.60 / **3.78**	2.37 / **3.32**	2.21 / **3.02**	2.09 / **2.80**	1.94 / **2.51**	1.75 / **2.18**	1.52 / **1.79**	

Table 3 is abridged from Table F of H. E. Garrett, *Statistics in Psychology and Education*, 5th edition, David McKay Co., Inc., New York, 1958, by permission of the author and publishers.

TABLE 4. CRITICAL VALUES OF *r*

df	P = .1000	P = .0500	P = .0200	P = .0100
1	.988	.997	.9995	.9999
2	.900	.950	.980	.990
3	.805	.878	.934	.959
4	.729	.811	.882	.917
5	.669	.754	.833	.874
6	.622	.707	.789	.834
7	.582	.666	.750	.798
8	.549	.632	.716	.765
9	.521	.602	.685	.735
10	.497	.576	.658	.708
11	.476	.553	.634	.684
12	.458	.532	.612	.661
13	.441	.514	.592	.641
14	.426	.497	.574	.623
15	.412	.482	.558	.606
16	.400	.468	.542	.590
17	.389	.456	.528	.575
18	.378	.444	.516	.561
19	.369	.433	.503	.549
20	.360	.423	.492	.537
21	.352	.413	.482	.526
22	.344	.404	.472	.515
23	.337	.396	.462	.505
24	.330	.388	.453	.496
25	.323	.381	.445	.487
26	.317	.374	.437	.479
27	.311	.367	.430	.471
28	.306	.361	.423	.463
29	.301	.355	.416	.456
30	.296	.349	.409	.449
35	.275	.325	.381	.418
40	.257	.304	.358	.393
45	.243	.288	.338	.372
50	.231	.273	.322	.354
60	.211	.250	.295	.325
70	.195	.232	.274	.302
80	.183	.217	.256	.283
90	.173	.205	.242	.267
100	.164	.195	.230	.254

Table 4 is taken from Table V.A. of Fisher: *Statistical Methods for Research Workers*, published by Oliver & Boyd Limited, Edinburgh, and by permission of the author and publishers.

TABLE 5. CRITICAL VALUES OF SPEARMAN'S RANK CORRELATION COEFFICIENT (rho)

N	$P = 0.10$	$P = 0.05$	$P = 0.02$	$P = 0.01$
5	0.900	—	—	—
6	0.829	0.886	0.943	—
7	0.714	0.786	0.893	0.929
8	0.643	0.738	0.833	0.881
9	0.600	0.683	0.783	0.833
10	0.564	0.648	0.745	0.794
11	0.523	0.623	0.736	0.818
12	0.497	0.591	0.703	0.780
13	0.475	0.566	0.673	0.745
14	0.457	0.545	0.646	0.716
15	0.441	0.525	0.623	0.689
16	0.425	0.507	0.601	0.666
17	0.412	0.490	0.582	0.645
18	0.399	0.476	0.564	0.625
19	0.388	0.462	0.549	0.608
20	0.377	0.450	0.534	0.591
21	0.368	0.438	0.521	0.576
22	0.359	0.428	0.508	0.562
23	0.351	0.418	0.496	0.549
24	0.343	0.409	0.485	0.537
25	0.336	0.400	0.475	0.526
26	0.329	0.392	0.465	0.515
27	0.323	0.385	0.456	0.505
28	0.317	0.377	0.448	0.496
29	0.311	0.370	0.440	0.487
30	0.305	0.364	0.432	0.478

Table 5 is taken from N. L. Johnson and F. C. Leone, *Statistical and Experimental Design, Vol. 1* (New York: John Wiley & Sons, Inc., 1964), by permission of the publisher.

TABLE 6.   CRITICAL VALUES OF CHI SQUARE (χ^2)

df	$P = .0500$	$P = .0100$
1	3.84	6.64
2	5.99	9.21
3	7.82	11.34
4	9.49	13.28
5	11.07	15.09
6	12.59	16.81
7	14.07	18.48
8	15.51	20.09
9	16.92	21.67
10	18.31	23.21
11	19.68	24.72
12	21.03	26.22
13	22.36	27.69
14	23.68	29.14
15	25.00	30.58
16	26.30	32.00
17	27.59	33.41
18	28.87	34.80
19	30.14	36.19
20	31.41	37.57
21	32.67	38.93
22	33.92	40.29
23	35.17	41.64
24	36.42	42.98
25	37.65	44.31
26	38.88	45.64
27	40.11	46.96
28	41.34	48.28
29	42.56	49.59
30	43.77	50.89

Table 6 is taken from Table 4 of Fisher & Yates, *Statistical Tables for Biological, Agricultural and Medical Research*, published by Oliver & Boyd Limited, Edinburgh, and by permission of the authors and publishers.

TABLE 7.   SQUARES AND SQUARE ROOTS

There are a few points to watch in using the table on the following pages. We shall discuss these briefly. First of all, observe that the first column lists all numbers, n, from 1.00 through 10.00. Each number in the second column is the square, n^2 of the corresponding number, n, in the first column. For example, $(1.78)^2 = 3.1684$ and $(7.17)^2 = 51.4089$.

The second column can also be used to obtain the squares of other numbers having the same succession of digits as the numbers given in the first column. For example, the square of 17.8 will also have the same succession of digits, 31684, as the square of 1.78. However, the position of the decimal point is not the same and $(17.8)^2 = 316.84$. This can be explained by the fact that we must multiply 1.78 by 10 to get 17.8. When the number is squared, the 10 is also squared $[(17.8)^2 = (10)^2 \cdot (1.78)^2]$. Thus the answer is 100 times 3.1684 or 316.84. Similarly, $(717)^2 = 514,089$ because 7.17 must be multiplied by 100 to give 717 and the square of 7.17 is then multiplied by the square of 100, which is 10,000. Note that in each case *the decimal point is moved twice as many places in the square as in the number that is squared.* Consider now the effect of moving the decimal point in the opposite direction. $(.178)^2$ will again contain the digits 31684 but this time the correct answer is .031684. The explanation is that 1.78 must be multiplied by .1 to give .178. The answer is then multiplied by $(.1)^2$ or .01 thus giving .031684. As another example, $(.0717)^2 = .00514089$. Note that the italicized statement holds regardless of the direction that the decimal point is moved.

The operation of taking the square root is the inverse of the operation of squaring, just as division is the inverse operation of multiplication. The discussion in the preceding paragraph is, therefore, also helpful in understanding the use of the third column of the table. This column gives the square root, $\sqrt{n}$, of the corresponding number, n, in the first column of the table. To further simplify the use of the table for finding square roots, the fourth column, $\sqrt{10n}$, has been added.

Since the first column contains all numbers from 1.00 to 10.00, we know that the third column contains the square roots of all these numbers. For example, $\sqrt{1.78} = 1.33417$ and $\sqrt{7.17} = 2.67769$. The fourth column enables us also to find directly the square roots, $\sqrt{10n}$, of all numbers from 10 (1.00) to 10(10.00), that is, from 10.0 to 100.0, where now each number is given only to the nearest tenth. For example, we find from the fourth column opposite 1.78 that $\sqrt{17.8} = 4.21900$ and opposite 7.17 that $\sqrt{71.7} = 8.46759$. Hence from the third and fourth columns we can read directly the square roots of all numbers from 1.00

From George Weinberg and John Schumaker, *Statistics: An Intuitive Approach, Second Edition* (Belmont, California: Brooks/Cole Publishing Company, 1969). Reprinted by permission of the authors and publisher.

through 100.0. However, just as we extended the use of the table for squares, so we can extend its use for square roots.

Suppose we want $\sqrt{717}$. From the table we can read both $\sqrt{7.17}$ and $\sqrt{71.7}$. Which should we use? This question is answered when we consider the placement of the decimal point. Remembering that taking the square root is the inverse of squaring, and looking back at the italicized statement earlier in the discussion, we see that *the decimal point is moved half as many places in the square root as in the number.* Now, half of an odd number isn't a whole number and a decimal point can't be moved a fraction of a place. So we must move the decimal point an *even number* of places to begin with when converting a number in order to apply the table and find its square root. Therefore, in our example we want the digits in $\sqrt{7.17}$ that are 267769 and by application of the last italicized statement we have $\sqrt{717} = 26.7769$, since the decimal point is moved two places from 7.17 to 717 and half of two is one. Had the problem been to find $\sqrt{7170}$, we would again have moved the decimal point an *even* number of places in order to obtain a number whose square root we could read directly from the table. In this case we would have the digits in $\sqrt{71.7}$ or 846759. Observing the rule for placement of the decimal point would give us the answer, 84.6759. The problem of finding $\sqrt{.00717}$ leads to the same sequence of digits (moving the decimal point four places), but the answer this time is .0846759. Again, the last italicized statement holds regardless of the direction that the decimal point is moved.

n	n^2	$\sqrt{n}$	$\sqrt{10n}$	n	n^2	$\sqrt{n}$	$\sqrt{10n}$
1.00	1.0000	1.00000	3.16228	1.50	2.2500	1.22474	3.87298
1.01	1.0201	1.00499	3.17805	1.51	2.2801	1.22882	3.88587
1.02	1.0404	1.00995	3.19374	1.52	2.3104	1.23288	3.89872
1.03	1.0609	1.01489	3.20936	1.53	2.3409	1.23693	3.91152
1.04	1.0816	1.01980	3.22490	1.54	2.3716	1.24097	3.92428
1.05	1.1025	1.02470	3.24037	1.55	2.4025	1.24499	3.93700
1.06	1.1236	1.02956	3.25576	1.56	2.4336	1.24900	3.94968
1.07	1.1449	1.03441	3.27109	1.57	2.4649	1.25300	3.96232
1.08	1.1664	1.03923	3.28634	1.58	2.4964	1.25698	3.97492
1.09	1.1881	1.04403	3.30151	1.59	2.5281	1.26095	3.98748
1.10	1.2100	1.04881	3.31662	1.60	2.5600	1.26491	4.00000
1.11	1.2321	1.05357	3.33167	1.61	2.5921	1.26886	4.01248
1.12	1.2544	1.05830	3.34664	1.62	2.6244	1.27279	4.02492
1.13	1.2769	1.06301	3.36155	1.63	2.6569	1.27671	4.03733
1.14	1.2996	1.06771	3.37639	1.64	2.6896	1.28062	4.04969
1.15	1.3225	1.07238	3.39116	1.65	2.7225	1.28452	4.06202
1.16	1.3456	1.07703	3.40588	1.66	2.7556	1.28841	4.07431
1.17	1.3689	1.08167	3.42053	1.67	2.7889	1.29228	4.08656
1.18	1.3924	1.08628	3.43511	1.68	2.8224	1.29615	4.09878
1.19	1.4161	1.09087	3.44964	1.69	2.8561	1.30000	4.11096
1.20	1.4400	1.09545	3.46410	1.70	2.8900	1.30384	4.12311
1.21	1.4641	1.10000	3.47851	1.71	2.9241	1.30767	4.13521
1.22	1.4884	1.10454	3.49285	1.72	2.9584	1.31149	4.14729
1.23	1.5129	1.10905	3.50714	1.73	2.9929	1.31529	4.15933
1.24	1.5376	1.11355	3.52136	1.74	3.0276	1.31909	4.17133
1.25	1.5625	1.11803	3.53553	1.75	3.0625	1.32288	4.18330
1.26	1.5876	1.12250	3.54965	1.76	3.0976	1.32665	4.19524
1.27	1.6129	1.12694	3.56371	1.77	3.1329	1.33041	4.20714
1.28	1.6384	1.13137	3.57771	1.78	3.1684	1.33417	4.21900
1.29	1.6641	1.13578	3.59166	1.79	3.2041	1.33791	4.23084
1.30	1.6900	1.14018	3.60555	1.80	3.2400	1.34164	4.24264
1.31	1.7161	1.14455	3.61939	1.81	3.2761	1.34536	4.25441
1.32	1.7424	1.14891	3.63318	1.82	3.3124	1.34907	4.26615
1.33	1.7689	1.15326	3.64692	1.83	3.3489	1.35277	4.27785
1.34	1.7956	1.15758	3.66060	1.84	3.3856	1.35647	4.28952
1.35	1.8225	1.16190	3.67423	1.85	3.4225	1.36015	4.30116
1.36	1.8496	1.16619	3.68782	1.86	3.4596	1.36382	4.31277
1.37	1.8769	1.17047	3.70135	1.87	3.4969	1.36748	4.32435
1.38	1.9044	1.17473	3.71484	1.88	3.5344	1.37113	4.33590
1.39	1.9321	1.17898	3.72827	1.89	3.5721	1.37477	4.34741
1.40	1.9600	1.18322	3.74166	1.90	3.6100	1.37840	4.35890
1.41	1.9881	1.18743	3.75500	1.91	3.6481	1.38203	4.37035
1.42	2.0164	1.19164	3.76829	1.92	3.6864	1.38564	4.38178
1.43	2.0449	1.19583	3.78153	1.93	3.7249	1.38924	4.39318
1.44	2.0736	1.20000	3.79473	1.94	3.7636	1.39284	4.40454
1.45	2.1025	1.20416	3.80789	1.95	3.8025	1.39642	4.41588
1.46	2.1316	1.20830	3.82099	1.96	3.8416	1.40000	4.42719
1.47	2.1609	1.21244	3.83406	1.97	3.8809	1.40357	4.43847
1.48	2.1904	1.21655	3.84708	1.98	3.9204	1.40712	4.44972
1.49	2.2201	1.22066	3.86005	1.99	3.9601	1.41067	4.46094

n	n^2	$\sqrt{n}$	$\sqrt{10n}$	n	n^2	$\sqrt{n}$	$\sqrt{10n}$
2.00	4.0000	1.41421	4.47214	2.50	6.2500	1.58114	5.00000
2.01	4.0401	1.41774	4.48330	2.51	6.3001	1.58430	5.00999
2.02	4.0804	1.42127	4.49444	2.52	6.3504	1.58745	5.01996
2.03	4.1209	1.42478	4.50555	2.53	6.4009	1.59060	5.02991
2.04	4.1616	1.42829	4.51664	2.54	6.4516	1.59374	5.03984
2.05	4.2025	1.43178	4.52769	2.55	6.5025	1.59687	5.04975
2.06	4.2436	1.43527	4.53872	2.56	6.5536	1.60000	5.05964
2.07	4.2849	1.43875	4.54973	2.57	6.6049	1.60312	5.06952
2.08	4.3264	1.44222	4.56070	2.58	6.6564	1.60624	5.07937
2.09	4.3681	1.44568	4.57165	2.59	6.7081	1.60935	5.08920
2.10	4.4100	1.44914	4.58258	2.60	6.7600	1.61245	5.09902
2.11	4.4521	1.45258	4.59347	2.61	6.8121	1.61555	5.10882
2.12	4.4944	1.45602	4.60435	2.62	6.8644	1.61864	5.11859
2.13	4.5369	1.45945	4.61519	2.63	6.9169	1.62173	5.12835
2.14	4.5796	1.46287	4.62601	2.64	6.9696	1.62481	5.13809
2.15	4.6225	1.46629	4.63681	2.65	7.0225	1.62788	5.14782
2.16	4.6656	1.46969	4.64758	2.66	7.0756	1.63095	5.15752
2.17	4.7089	1.47309	4.65833	2.67	7.1289	1.63401	5.16720
2.18	4.7524	1.47648	4.66905	2.68	7.1824	1.63707	5.17687
2.19	4.7961	1.47986	4.67974	2.69	7.2361	1.64012	5.18652
2.20	4.8400	1.48324	4.69042	2.70	7.2900	1.64317	5.19615
2.21	4.8841	1.48661	4.70106	2.71	7.3441	1.64621	5.20577
2.22	4.9284	1.48997	4.71169	2.72	7.3984	1.64924	5.21536
2.23	4.9729	1.49332	4.72229	2.73	7.4529	1.65227	5.22494
2.24	5.0176	1.49666	4.73286	2.74	7.5076	1.65529	5.23450
2.25	5.0625	1.50000	4.74342	2.75	7.5625	1.65831	5.24404
2.26	5.1076	1.50333	4.75395	2.76	7.6176	1.66132	5.25357
2.27	5.1529	1.50665	4.76445	2.77	7.6729	1.66433	5.26308
2.28	5.1984	1.50997	4.77493	2.78	7.7284	1.66733	5.27257
2.29	5.2441	1.51327	4.78539	2.79	7.7841	1.67033	5.28205
2.30	5.2900	1.51658	4.79583	2.80	7.8400	1.67332	5.29150
2.31	5.3361	1.51987	4.80625	2.81	7.8961	1.67631	5.30094
2.32	5.3824	1.52315	4.81664	2.82	7.9524	1.67929	5.31037
2.33	5.4289	1.52643	4.82701	2.83	8.0089	1.68226	5.31977
2.34	5.4756	1.52971	4.83735	2.84	8.0656	1.68523	5.32917
2.35	5.5225	1.53297	4.84768	2.85	8.1225	1.68819	5.33854
2.36	5.5696	1.53623	4.85798	2.86	8.1796	1.69115	5.34790
2.37	5.6169	1.53948	4.86826	2.87	8.2369	1.69411	5.35724
2.38	5.6644	1.54272	4.87852	2.88	8.2944	1.69706	5.36656
2.39	5.7121	1.54596	4.88876	2.89	8.3521	1.70000	5.37587
2.40	5.7600	1.54919	4.89898	2.90	8.4100	1.70294	5.38516
2.41	5.8081	1.55242	4.90918	2.91	8.4681	1.70587	5.39444
2.42	5.8564	1.55563	4.91935	2.92	8.5264	1.70880	5.40370
2.43	5.9049	1.55885	4.92950	2.93	8.5849	1.71172	5.41295
2.44	5.9536	1.56205	4.93964	2.94	8.6436	1.71464	5.42218
2.45	6.0025	1.56525	4.94975	2.95	8.7025	1.71756	5.43139
2.46	6.0516	1.56844	4.95984	2.96	8.7616	1.72047	5.44059
2.47	6.1009	1.57162	4.96991	2.97	8.8209	1.72337	5.44977
2.48	6.1504	1.57480	4.97996	2.98	8.8804	1.72627	5.45894
2.49	6.2001	1.57797	4.98999	2.99	8.9401	1.72916	5.46809

n	n^2	$\sqrt{n}$	$\sqrt{10n}$	n	n^2	$\sqrt{n}$	$\sqrt{10n}$
3.00	9.0000	1.73205	5.47723	3.50	12.2500	1.87083	5.91608
3.01	9.0601	1.73494	5.48635	3.51	12.3201	1.87350	5.92453
3.02	9.1204	1.73781	5.49545	3.52	12.3904	1.87617	5.93296
3.03	9.1809	1.74069	5.50454	3.53	12.4609	1.87883	5.94138
3.04	9.2416	1.74356	5.51362	3.54	12.5316	1.88149	5.94979
3.05	9.3025	1.74642	5.52268	3.55	12.6025	1.88414	5.95819
3.06	9.3636	1.74929	5.53173	3.56	12.6736	1.88680	5.96657
3.07	9.4249	1.75214	5.54076	3.57	12.7449	1.88944	5.97495
3.08	9.4864	1.75499	5.54977	3.58	12.8164	1.89209	5.98331
3.09	9.5481	1.75784	5.55878	3.59	12.8881	1.89473	5.99166
3.10	9.6100	1.76068	5.56776	3.60	12.9600	1.89737	6.00000
3.11	9.6721	1.76352	5.57674	3.61	13.0321	1.90000	6.00833
3.12	9.7344	1.76635	5.58570	3.62	13.1044	1.90263	6.01664
3.13	9.7969	1.76918	5.59464	3.63	13.1769	1.90526	6.02495
3.14	9.8596	1.77200	5.60357	3.64	13.2496	1.90788	6.03324
3.15	9.9225	1.77482	5.61249	3.65	13.3225	1.91050	6.04152
3.16	9.9856	1.77764	5.62139	3.66	13.3956	1.91311	6.04979
3.17	10.0489	1.78045	5.63028	3.67	13.4689	1.91572	6.05805
3.18	10.1124	1.78326	5.63915	3.68	13.5424	1.91833	6.06630
3.19	10.1761	1.78606	5.64801	3.69	13.6161	1.92094	6.07454
3.20	10.2400	1.78885	5.65685	3.70	13.6900	1.92354	6.08276
3.21	10.3041	1.79165	5.66569	3.71	13.7641	1.92614	6.09098
3.22	10.3684	1.79444	5.67450	3.72	13.8384	1.92873	6.09918
3.23	10.4329	1.79722	5.68331	3.73	13.9129	1.93132	6.10737
3.24	10.4976	1.80000	5.69210	3.74	13.9876	1.93391	6.11555
3.25	10.5625	1.80278	5.70088	3.75	14.0625	1.93649	6.12372
3.26	10.6276	1.80555	5.70964	3.76	14.1376	1.93907	6.13188
3.27	10.6929	1.80831	5.71839	3.77	14.2129	1.94165	6.14003
3.28	10.7584	1.81108	5.72713	3.78	14.2884	1.94422	6.14817
3.29	10.8241	1.81384	5.73585	3.79	14.3641	1.94679	6.15630
3.30	10.8900	1.81659	5.74456	3.80	14.4400	1.94936	6.16441
3.31	10.9561	1.81934	5.75326	3.81	14.5161	1.95192	6.17252
3.32	11.0224	1.82209	5.76194	3.82	14.5924	1.95448	6.18061
3.33	11.0889	1.82483	5.77062	3.83	14.6689	1.95704	6.18870
3.34	11.1556	1.82757	5.77927	3.84	14.7456	1.95959	6.19677
3.35	11.2225	1.83030	5.78792	3.85	14.8225	1.96214	6.20484
3.36	11.2896	1.83303	5.79655	3.86	14.8996	1.96469	6.21289
3.37	11.3569	1.83576	5.80517	3.87	14.9769	1.96723	6.22093
3.38	11.4244	1.83848	5.81378	3.88	15.0544	1.96977	6.22896
3.39	11.4921	1.84120	5.82237	3.89	15.1321	1.97231	6.23699
3.40	11.5600	1.84391	5.83095	3.90	15.2100	1.97484	6.24500
3.41	11.6281	1.84662	5.83952	3.91	15.2881	1.97737	6.25300
3.42	11.6964	1.84932	5.84808	3.92	15.3664	1.97990	6.26099
3.43	11.7649	1 85203	5 85662	3.93	15.4449	1.98242	6.26897
3.44	11.8336	1.85472	5.86515	3.94	15.5236	1.98494	6.27694
3.45	11.9025	1.85742	5.87367	3.95	15.6025	1.98746	6.28490
3.46	11.9716	1.86011	5.88218	3.96	15.6816	1.98997	6.29285
3.47	12.0409	1.86279	5.89067	3.97	15.7609	1.99249	6.30079
3.48	12.1104	1.86548	5.89915	3.98	15.8408	1.99499	6.30872
3.49	12.1801	1.86815	5.90762	3.99	15.9201	1.99750	6.31664

n	n^2	$\sqrt{n}$	$\sqrt{10n}$	n	n^2	$\sqrt{n}$	$\sqrt{10n}$
4.00	16.0000	2.00000	6.32456	4.50	20.2500	2.12132	6.70820
4.01	16.0801	2.00250	6.33246	4.51	20.3401	2.12368	6.71565
4.02	16.1604	2.00499	6.34035	4.52	20.4304	2.12603	6.72309
4.03	16.2409	2.00749	6.34823	4.53	20.5209	2.12838	6.73053
4.04	16.3216	2.00998	6.35610	4.54	20.6116	2.13073	6.73795
4.05	16.4025	2.01246	6.36396	4.55	20.7025	2.13307	6.74537
4.06	16.4836	2.01494	6.37181	4.56	20.7936	2.13542	6.75278
4.07	16.5649	2.01742	6.37966	4.57	20.8849	2.13776	6.76018
4.08	16.6464	2.01990	6.38749	4.58	20.9764	2.14009	6.76757
4.09	16.7281	2.02237	6.39531	4.59	21.0681	2.14243	6.77495
4.10	16.8100	2.02485	6.40312	4.60	21.1600	2.14476	6.78233
4.11	16.8921	2.02731	6.41093	4.61	21.2521	2.14709	6.78970
4.12	16.9744	2.02978	6.41872	4.62	21.3444	2.14942	6.79706
4.13	17.0569	2.03224	6.42651	4.63	21.4369	2.15174	6.80441
4.14	17.1396	2.03470	6.43428	4.64	21.5296	2.15407	6.81175
4.15	17.2225	2.03715	6.44205	4.65	21.6225	2.15639	6.81909
4.16	17.3056	2.03961	6.44981	4.66	21.7156	2.15870	6.82642
4.17	17.3889	2.04206	6.45755	4.67	21.8089	2.16102	6.83374
4.18	17.4724	2.04450	6.46529	4.68	21.9024	2.16333	6.84105
4.19	17.5561	2.04695	6.47302	4.69	21.9961	2.16564	6.84836
4.20	17.6400	2.04939	6.48074	4.70	22.0900	2.16795	6.85565
4.21	17.7241	2.05183	6.48845	4.71	22.1841	2.17025	6.86294
4.22	17.8084	2.05426	6.49615	4.72	22.2784	2.17256	6.87023
4.23	17.8929	2.05670	6.50384	4.73	22.3729	2.17486	6.87750
4.24	17.9776	2.05913	6.51153	4.74	22.4676	2.17715	6.88477
4.25	18.0625	2.06155	6.51920	4.75	22.5625	2.17945	6.89202
4.26	18.1476	2.06398	6.52687	4.76	22.6576	2.18174	6.89928
4.27	18.2329	2.06640	6.53452	4.77	22.7529	2.18403	6.90652
4.28	18.3184	2.06882	6.54217	4.78	22.8484	2.18632	6.91375
4.29	18.4041	2.07123	6.54981	4.79	22.9441	2.18861	6.92098
4.30	18.4900	2.07364	6.55744	4.80	23.0400	2.19089	6.92820
4.31	18.5761	2.07605	6.56506	4.81	23.1361	2.19317	6.93542
4.32	18.6624	2.07846	6.57267	4.82	23.2324	2.19545	6.94262
4.33	18.7489	2.08087	6.58027	4.83	23.3289	2.19773	6.94982
4.34	18.8356	2.08327	6.58787	4.84	23.4256	2.20000	6.95701
4.35	18.9225	2.08567	6.59545	4.85	23.5225	2.20227	6.96419
4.36	19.0096	2.08806	6.60303	4.86	23.6196	2.20454	6.97137
4.37	19.0969	2.09045	6.61060	4.87	23.7169	2.20681	6.97854
4.38	19.1844	2.09284	6.61816	4.88	23.8144	2.20907	6.98570
4.39	19.2721	2.09523	6.62571	4.89	23.9121	2.21133	6.99285
4.40	19.3600	2.09762	6.63325	4.90	24.0100	2.21359	7.00000
4.41	19.4481	2.10000	6.64078	4.91	24.1081	2.21585	7.00714
4.42	19.5364	2.10238	6.64831	4.92	24.2064	2.21811	7.01427
4.43	19.6249	2.10476	6.65582	4.93	24.3049	2.22036	7.02140
4.44	19.7136	2.10713	6.66333	4.94	24.4036	2.22261	7.02851
4.45	19.8025	2.10950	6.67083	4.95	24.5025	2.22486	7.03562
4.46	19.8916	2.11187	6.67832	4.96	24.6016	2.22711	7.04273
4.47	19.9809	2.11424	6.68581	4.97	24.7009	2.22935	7.04982
4.48	20.0704	2.11660	6.69328	4.98	24.8004	2.23159	7.05691
4.49	20.1601	2.11896	6.70075	4.99	24.9001	2.23383	7.06399

n	n^2	$\sqrt{n}$	$\sqrt{10n}$	n	n^2	$\sqrt{n}$	$\sqrt{10n}$
5.00	25.0000	2.23607	7.07107	5.50	30.2500	2.34521	7.41620
5.01	25.1001	2.23830	7.07814	5.51	30.3601	2.34734	7.42294
5.02	25.2004	2.24054	7.08520	5.52	30.4704	2.34947	7.42967
5.03	25.3009	2.24277	7.09225	5.53	30.5809	2.35160	7.43640
5.04	25.4016	2.24499	7.09930	5.54	30.6916	2.35372	7.44312
5.05	25.5025	2.24722	7.10634	5.55	30.8025	2.35584	7.44983
5.06	25.6036	2.24944	7.11337	5.56	30.9136	2.35797	7.45654
5.07	25.7049	2.25167	7.12039	5.57	31.0249	2.36008	7.46324
5.08	25.8064	2.25389	7.12741	5.58	31.1364	2.36220	7.46994
5.09	25.9081	2.25610	7.13442	5.59	31.2481	2.36432	7.47663
5.10	26.0100	2.25832	7.14143	5.60	31.3600	2.36643	7.48331
5.11	26.1121	2.26053	7.14843	5.61	31.4721	2.36854	7.48999
5.12	26.2144	2.26274	7.15542	5.62	31.5844	2.37065	7.49667
5.13	26.3169	2.26495	7.16240	5.63	31.6969	2.37276	7.50333
5.14	26.4196	2.26716	7.16938	5.64	31.8096	2.37487	7.50999
5.15	26.5225	2.26936	7.17635	5.65	31.9225	2.37697	7.51665
5.16	26.6256	2.27156	7.18331	5.66	32.0356	2.37908	7.52330
5.17	26.7289	2.27376	7.19027	5.67	32.1489	2.38118	7.52994
5.18	26.8324	2.27596	7.19722	5.68	32.2624	2.38328	7.53658
5.19	26.9361	2.27816	7.20417	5.69	32.3761	2.38537	7.54321
5.20	27.0400	2.28035	7.21110	5.70	32.4900	2.38747	7.54983
5.21	27.1441	2.28254	7.21803	5.71	32.6041	2.38956	7.55645
5.22	27.2484	2.28473	7.22496	5.72	32.7184	2.39165	7.56307
5.23	27.3529	2.28692	7.23187	5.73	32.8329	2.39374	7.56968
5.24	27.4576	2.28910	7.23878	5.74	32.9476	2.39583	7.57628
5.25	27.5625	2.29129	7.24569	5.75	33.0625	2.39792	7.58288
5.26	27.6676	2.29347	7.25259	5.76	33.1776	2.40000	7.58947
5.27	27.7729	2.29565	7.25948	5.77	33.2929	2.40208	7.59605
5.28	27.8784	2.29783	7.26636	5.78	33.4084	2.40416	7.60263
5.29	27.9841	2.30000	7.27324	5.79	33.5241	2.40624	7.60920
5.30	28.0900	2.30217	7.28011	5.80	33.6400	2.40832	7.61577
5.31	28.1961	2.30434	7.28697	5.81	33.7561	2.41039	7.62234
5.32	28.3024	2.30651	7.29383	5.82	33.8724	2.41247	7.62889
5.33	28.4089	2.30868	7.30068	5.83	33.9889	2.41454	7.63544
5.34	28.5156	2.31084	7.30753	5.84	34.1056	2.41661	7.64199
5.35	28.6225	2.31301	7.31437	5.85	34.2225	2.41868	7.64853
5.36	28.7296	2.31517	7.32120	5.86	34.3396	2.42074	7.65506
5.37	28.8369	2.31733	7.32803	5.87	34.4569	2.42281	7.66159
5.38	28.9444	2.31948	7.33485	5.88	34.5744	2.42487	7.66812
5.39	29.0521	2.32164	7.34166	5.89	34.6921	2.42693	7.67463
5.40	29.1600	2.32379	7.34847	5.90	34.8100	2.42899	7.68115
5.41	29.2681	2.32594	7.35527	5.91	34.9281	2.43105	7.68765
5.42	29.3764	2.32809	7.36205	5.92	35.0464	2.43311	7.69415
5.43	29.4849	2.33024	7.36885	5.93	35.1649	2.43516	7.70065
5.44	29.5936	2.33238	7.37564	5.94	35.2836	2.43721	7.70714
5.45	29.7025	2.33452	7.38241	5.95	35.4025	2.43926	7.71362
5.46	29.8116	2.33666	7.38918	5.96	35.5216	2.44131	7.72010
5.47	29.9209	2.33880	7.39594	5.97	35.6409	2.44336	7.72658
5.48	30.0304	2.34094	7.40270	5.98	35.7604	2.44540	7.73305
5.49	30.1401	2.34307	7.40945	5.99	35.8801	2.44745	7.73951

374

n	n^2	$\sqrt{n}$	$\sqrt{10n}$	n	n^2	$\sqrt{n}$	$\sqrt{10n}$
6.00	36.0000	2.44949	7.74597	6.50	42.2500	2.54951	8.06226
6.01	36.1201	2.45153	7.75242	6.51	42.3801	2.55147	8.06846
6.02	36.2404	2.45357	7.75887	6.52	42.5104	2.55343	8.07465
6.03	36.3609	2.45561	7.76531	6.53	42.6409	2.55539	8.08084
6.04	36.4816	2.45764	7.77174	6.54	42.7716	2.55734	8.08703
6.05	36.6025	2.45967	7.77817	6.55	42.9025	2.55930	8.09321
6.06	36.7236	2.46171	7.78460	6.56	43.0336	2.56125	8.09938
6.07	36.8449	2.46374	7.79102	6.57	43.1649	2.56320	8.10555
6.08	36.9664	2.46577	7.79744	6.58	43.2964	2.56515	8.11172
6.09	37.0881	2.46779	7.80385	6.59	43.4281	2.56710	8.11788
6.10	37.2100	2.46982	7.81025	6.60	43.5600	2.56905	8.12404
6.11	37.3321	2.47184	7.81665	6.61	43.6921	2.57099	8.13019
6.12	37.4544	2.47386	7.82304	6.62	43.8244	2.57294	8.13634
6.13	37.5769	2.47588	7.82943	6.63	43.9569	2.57488	8.14248
6.14	37.6996	2.47790	7.83582	6.64	44.0896	2.57682	8.14862
6.15	37.8225	2.47992	7.84219	6.65	44.2225	2.57876	8.15475
6.16	37.9456	2.48193	7.84857	6.66	44.3556	2.58070	8.16088
6.17	38.0689	2.48395	7.85493	6.67	44.4889	2.58263	8.16701
6.18	38.1924	2.48596	7.86130	6.68	44.6224	2.58457	8.17313
6.19	38.3161	2.48797	7.86766	6.69	44.7561	2.58650	8.17924
6.20	38.4400	2.48998	7.87401	6.70	44.8900	2.58844	8.18535
6.21	38.5641	2.49199	7.88036	6.71	45.0241	2.59037	8.19146
6.22	38.6884	2.49399	7.88670	6.72	45.1584	2.59230	8.19756
6.23	38.8129	2.49600	7.89303	6.73	45.2929	2.59422	8.20366
6.24	38.9376	2.49800	7.89937	6.74	45.4276	2.59615	8.20975
6.25	39.0625	2.50000	7.90569	6.75	45.5625	2.59808	8.21584
6.26	39.1876	2.50200	7.91202	6.76	45.6976	2.60000	8.22192
6.27	39.3129	2.50400	7.91833	6.77	45.8329	2.60192	8.22800
6.28	39.4384	2.50599	7.92465	6.78	45.9684	2.60384	8.23408
6.29	39.5641	2.50799	7.93095	6.79	46.1041	2.60576	8.24015
6.30	39.6900	2.50998	7.93725	6.80	46.2400	2.60768	8.24621
6.31	39.8161	2.51197	7.94355	6.81	46.3761	2.60960	8.25227
6.32	39.9424	2.51396	7.94984	6.82	46.5124	2.61151	8.25833
6.33	40.0689	2.51595	7.95613	6.83	46.6489	2.61343	8.26438
6.34	40.1956	2.51794	7.96241	6.84	46.7856	2.61534	8.27043
6.35	40.3225	2.51992	7.96869	6.85	46.9225	2.61725	8.27647
6.36	40.4496	2.52190	7.97496	6.86	47.0596	2.61916	8.28251
6.37	40.5769	2.52389	7.98123	6.87	47.1969	2.62107	8.28855
6.38	40.7044	2.52587	7.98749	6.88	47.3344	2.62298	8.29458
6.39	40.8321	2.52784	7.99375	6.89	47.4721	2.62488	8.30060
6.40	40.9600	2.52982	8.00000	6.90	47.6100	2.62679	8.30662
6.41	41.0881	2.53180	8.00625	6.91	47.7481	2.62869	8.31264
6.42	41.2164	2.53377	8.01249	6.92	47.8864	2.63059	8.31865
6.43	41.3449	2.53574	8.01873	6.93	48.0249	2.63249	8.32466
6.44	41.4736	2.53772	8.02496	6.94	48.1636	2.63439	8.33067
6.45	41.6025	2.53969	8.03119	6.95	48.3025	2.63629	8.33667
6.46	41.7316	2.54165	8.03741	6.96	48.4416	2.63818	8.34266
6.47	41.8609	2.54362	8.04363	6.97	48.5809	2.64008	8.34865
6.48	41.9904	2.54558	8.04984	6.98	48.7204	2.64197	8.35464
6.49	42.1201	2.54755	8.05605	6.99	48.8601	2.64386	8.36062

n	n^2	$\sqrt{n}$	$\sqrt{10n}$	n	n^2	$\sqrt{n}$	$\sqrt{10n}$
7.00	49.0000	2.64575	8.36660	7.50	56.2500	2.73861	8.66025
7.01	49.1401	2.64764	8.37257	7.51	56.4001	2.74044	8.66603
7.02	49.2804	2.64953	8.37854	7.52	56.5504	2.74226	8.67179
7.03	49.4209	2.65141	8.38451	7.53	56.7009	2.74408	8.67756
7.04	49.5616	2.65330	8.39047	7.54	56.8516	2.74591	8.68332
7.05	49.7025	2.65518	8.39643	7.55	57.0025	2.74773	8.68907
7.06	49.8436	2.65707	8.40238	7.56	57.1536	2.74955	8.69483
7.07	49.9849	2.65895	8.40833	7.57	57.3049	2.75136	8.70057
7.08	50.1264	2.66083	8.41427	7.58	57.4564	2.75318	8.70632
7.09	50.2681	2.66271	8.42021	7.59	57.6081	2.75500	8.71206
7.10	50.4100	2.66458	8.42615	7.60	57.7600	2.75681	8.71780
7.11	50.5521	2.66646	8.43208	7.61	57.9121	2.75862	8.72353
7.12	50.6944	2.66833	8.43801	7.62	58.0644	2.76043	8.72926
7.13	50.8369	2.67021	8.44393	7.63	58.2169	2.76225	8.73499
7.14	50.9796	2.67208	8.44985	7.64	58.3696	2.76405	8.74071
7.15	51.1225	2.67395	8.45577	7.65	58.5225	2.76586	8.74643
7.16	51.2656	2.67582	8.46168	7.66	58.6756	2.76767	8.75214
7.17	51.4089	2.67769	8.46759	7.67	58.8289	2.76948	8.75785
7.18	51.5524	2.67955	8.47349	7.68	58.9824	2.77128	8.76356
7.19	51.6961	2.68142	8.47939	7.69	59.1361	2.77308	8.76926
7.20	51.8400	2.68328	8.48528	7.70	59.2900	2.77489	8.77496
7.21	51.9841	2.68514	8.49117	7.71	59.4441	2.77669	8.78066
7.22	52.1284	2.68701	8.49706	7.72	59.5984	2.77849	8.78635
7.23	52.2729	2.68887	8.50294	7.73	59.7529	2.78029	8.79204
7.24	52.4176	2.69072	8.50882	7.74	59.9076	2.78209	8.79773
7.25	52.5625	2.69258	8.51469	7.75	60.0625	2.78388	8.80341
7.26	52.7076	2.69444	8.52056	7.76	60.2176	2.78568	8.80909
7.27	52.8529	2.69629	8.52643	7.77	60.3729	2.78747	8.81476
7.28	52.9984	2.69815	8.53229	7.78	60.5284	2.78927	8.82043
7.29	53.1441	2.70000	8.53815	7.79	60.6841	2.79106	8.82610
7.30	53.2900	2.70185	8.54400	7.80	60.8400	2.79285	8.83176
7.31	53.4361	2.70370	8.54985	7.81	60.9961	2.79464	8.83742
7.32	53.5824	2.70555	8.55570	7.82	61.1524	2.79643	8.84308
7.33	53.7289	2.70740	8.56154	7.83	61.3089	2.79821	8.84873
7.34	53.8756	2.70924	8.56738	7.84	61.4656	2.80000	8.85438
7.35	54.0225	2.71109	8.57321	7.85	61.6225	2.80179	8.86002
7.36	54.1696	2.71293	8.57904	7.86	61.7796	2.80357	8.86566
7.37	54.3169	2.71477	8.58487	7.87	61.9369	2.80535	8.87130
7.38	54.4644	2.71662	8.59069	7.88	62.0944	2.80713	8.87694
7.39	54.6121	2.71846	8.59651	7.89	62.2521	2.80891	8.88257
7.40	54.7600	2.72029	8.60233	7.90	62.4100	2.81069	8.88819
7.41	54.9081	2.72213	8.60814	7.91	62.5681	2.81247	8.89382
7.42	55.0564	2.72397	8.61394	7.92	62.7264	2.81425	8.89944
7.43	55.2049	2.72580	8.61974	7.93	62.8849	2.81603	8.90505
7.44	55.3536	2.72764	8.62554	7.94	63.0436	2.81780	8.91067
7.45	55.5025	2.72947	8.63134	7.95	63.2025	2.81957	8.91628
7.46	55.6516	2.73130	8.63713	7.96	63.3616	2.82135	8.92188
7.47	55.8009	2.73313	8.64292	7.97	63.5209	2.82312	8.92749
7.48	55.9504	2.73496	8.64870	7.98	63.6804	2.82489	8.93308
7.49	56.1001	2.73679	8.65448	7.99	63.8401	2.82666	8.93868

n	n^2	$\sqrt{n}$	$\sqrt{10n}$	n	n^2	$\sqrt{n}$	$\sqrt{10n}$
8.00	64.0000	2.82843	8.94427	8.50	72.2500	2.91548	9.21954
8.01	64.1601	2.83019	8.94986	8.51	72.4201	2.91719	9.22497
8.02	64.3204	2.83196	8.95545	8.52	72.5904	2.91890	9.23038
8.03	64.4809	2.83373	8.96103	8.53	72.7609	2.92062	9.23580
8.04	64.6416	2.83549	8.96660	8.54	72.9316	2.92233	9.24121
8.05	64.8025	2.83725	8.97218	8.55	73.1025	2.92404	9.24662
8.06	64.9636	2.83901	8.97775	8.56	73.2736	2.92575	9.25203
8.07	65.1249	2.84077	8.98332	8.57	73.4449	2.92746	9.25743
8.08	65.2864	2.84253	8.98888	8.58	73.6164	2.92916	9.26283
8.09	65.4481	2.84429	8.99444	8.59	73.7881	2.93087	9.26823
8.10	65.6100	2.84605	9.00000	8.60	73.9600	2.93258	9.27362
8.11	65.7721	2.84781	9.00555	8.61	74.1321	2.93428	9.27901
8.12	65.9344	2.84956	9.01110	8.62	74.3044	2.93598	9.28440
8.13	66.0969	2.85132	9.01665	8.63	74.4769	2.93769	9.28978
8.14	66.2596	2.85307	9.02219	8.64	74.6496	2.93939	9.29516
8.15	66.4225	2.85482	9.02774	8.65	74.8225	2.94109	9.30054
8.16	66.5856	2.85657	9.03327	8.66	74.9956	2.94279	9.30591
8.17	66.7489	2.85832	9.03881	8.67	75.1689	2.94449	9.31128
8.18	66.9124	2.86007	9.04434	8.68	75.3424	2.94618	9.31665
8.19	67.0761	2.86182	9.04986	8.69	75.5161	2.94788	9.32202
8.20	67.2400	2.86356	9.05539	8.70	75.6900	2.94958	9.32738
8.21	67.4041	2.86531	9.06091	8.71	75.8641	2.95127	9.33274
8.22	67.5684	2.86705	9.06642	8.72	76.0384	2.95296	9.33809
8.23	67.7329	2.86880	9.07193	8.73	76.2129	2.95466	9.34345
8.24	67.8976	2.87054	9.07744	8.74	76.3876	2.95635	9.34880
8.25	68.0625	2.87228	9.08295	8.75	76.5625	2.95804	9.35414
8.26	68.2276	2.87402	9.08845	8.76	76.7376	2.95973	9.35949
8.27	68.3929	2.87576	9.09395	8.77	76.9129	2.96142	9.36483
8.28	68.5584	2.87750	9.09945	8.78	77.0884	2.96311	9.37017
8.29	68.7241	2.87924	9.10494	8.79	77.2641	2.96479	9.37550
8.30	68.8900	2.88097	9.11043	8.80	77.4400	2.96648	9.38083
8.31	69.0561	2.88271	9.11592	8.81	77.6161	2.96816	9.38616
8.32	69.2224	2.88444	9.12140	8.82	77.7924	2.96985	9.39149
8.33	69.3889	2.88617	9.12688	8.83	77.9689	2.97153	9.39681
8.34	69.5556	2.88791	9.13236	8.84	78.1456	2.97321	9.40213
8.35	69.7225	2.88964	9.13783	8.85	78.3225	2.97489	9.40744
8.36	69.8896	2.89137	9.14330	8.86	78.4996	2.97658	9.41276
8.37	70.0569	2.89310	9.14877	8.87	78.6769	2.97825	9.41807
8.38	70.2244	2.89482	9.15423	8.88	78.8544	2.97993	9.42338
8.39	70.3921	2.89655	9.15969	8.89	79.0321	2.98161	9.42868
8.40	70.5600	2.89828	9.16515	8.90	79.2100	2.98329	9.43398
8.41	70.7281	2.90000	9.17061	8.91	79.3881	2.98496	9.43928
8.42	70.8964	2.90172	9.17606	8.92	79.5664	2.98664	9.44458
8.43	71.0649	2.90345	9.18150	8.93	79.7449	2.98831	9.44987
8.44	71.2336	2.90517	9.18695	8.94	79.9236	2.98998	9.45516
8.45	71.4025	2.90689	9.19239	8.95	80.1025	2.99166	9.46044
8.46	71.5716	2.90861	9.19783	8.96	80.2816	2.99333	9.46573
8.47	71.7409	2.91033	9.20326	8.97	80.4609	2.99500	9.47101
8.48	71.9104	2.91204	9.20869	8.98	80.6404	2.99666	9.47629
8.49	72.0801	2.91376	9.21412	8.99	80.8201	2.99833	9.48156

n	n^2	$\sqrt{n}$	$\sqrt{10n}$	n	n^2	$\sqrt{n}$	$\sqrt{10n}$
9.00	81.0000	3.00000	9.48683	9.50	90.2500	3.08221	9.74679
9.01	81.1801	3.00167	9.49210	9.51	90.4401	3.08383	9.75192
9.02	81.3604	3.00333	9.49737	9.52	90.6304	3.08545	9.75705
9.03	81.5409	3.00500	9.50263	9.53	90.8209	3.08707	9.76217
9.04	81.7216	3.00666	9.50789	9.54	91.0116	3.08869	9.76729
9.05	81.9025	3.00832	9.51315	9.55	91.2025	3.09031	9.77241
9.06	82.0836	3.00998	9.51840	9.56	91.3936	3.09192	9.77753
9.07	82.2649	3.01164	9.52365	9.57	91.5849	3.09354	9.78264
9.08	82.4464	3.01330	9.52890	9.58	91.7764	3.09516	9.78775
9.09	82.6281	3.01496	9.53415	9.59	91.9681	3.09677	9.79285
9.10	82.8100	3.01662	9.53939	9.60	92.1600	3.09839	9.79796
9.11	82.9921	3.01828	9.54463	9.61	92.3521	3.10000	9.80306
9.12	83.1744	3.01993	9.54987	9.62	92.5444	3.10161	9.80816
9.13	83.3569	3.02159	9.55510	9.63	92.7369	3.10322	9.81326
9.14	83.5396	3.02324	9.56033	9.64	92.9296	3.10483	9.81835
9.15	83.7225	3.02490	9.56556	9.65	93.1225	3.10644	9.82344
9.16	83.9056	3.02655	9.57079	9.66	93.3156	3.10805	9.82853
9.17	84.0889	3.02820	9.57601	9.67	93.5089	3.10966	9.83362
9.18	84.2724	3.02985	9.58123	9.68	93.7024	3.11127	9.83870
9.19	84.4561	3.03150	9.58645	9.69	93.8961	3.11288	9.84378
9.20	84.6400	3.03315	9.59166	9.70	94.0900	3.11448	9.84886
9.21	84.8241	3.03480	9.59687	9.71	94.2841	3.11609	9.85393
9.22	85.0084	3.03645	9.60208	9.72	94.4784	3.11769	9.85901
9.23	85.1929	3.03809	9.60729	9.73	94.6729	3.11929	9.86408
9.24	85.3776	3.03974	9.61249	9.74	94.8676	3.12090	9.86914
9.25	85.5625	3.04138	9.61769	9.75	95.0625	3.12250	9.87421
9.26	85.7476	3.04302	9.62289	9.76	95.2576	3.12410	9.87927
9.27	85.9329	3.04467	9.62808	9.77	95.4529	3.12570	9.88433
9.28	86.1184	3.04631	9.63328	9.78	95.6484	3.12730	9.88939
9.29	86.3041	3.04795	9.63846	9.79	95.8441	3.12890	9.89444
9.30	86.4900	3.04959	9.64365	9.80	96.0400	3.13050	9.89949
9.31	86.6761	3.05123	9.64883	9.81	96.2361	3.13209	9.90454
9.32	86.8624	3.05287	9.65401	9.82	96.4324	3.13369	9.90959
9.33	87.0489	3.05450	9.65919	9.83	96.6289	3.13528	9.91464
9.34	87.2356	3.05614	9.66437	9.84	96.8256	3.13688	9.91968
9.35	87.4225	3.05778	9.66954	9.85	97.0225	3.13847	9.92472
9.36	87.6096	3.05941	9.67471	9.86	97.2196	3.14006	9.92975
9.37	87.7969	3.06105	9.67988	9.87	97.4169	3.14166	9.93479
9.38	87.9844	3.06268	9.68504	9.88	97.6144	3.14325	9.93982
9.39	88.1721	3.06431	9.69020	9.89	97.8121	3.14484	9.94485
9.40	88.3600	3.06594	9.69536	9.90	98.0100	3.14643	9.94987
9.41	88.5481	3.06757	9.70052	9.91	98.2081	3.14802	9.95490
9.42	88.7364	3.06920	9.70567	9.92	98.4064	3.14960	9.95992
9.43	88.9249	3.07083	9.71082	9.93	98.6049	3.15119	9.96494
9.44	89.1136	3.07246	9.71597	9.94	98.8036	3.15278	9.96995
9.45	89.3025	3.07409	9.72111	9.95	99.0025	3.15436	9.97497
9.46	89.4916	3.07571	9.72625	9.96	99.2016	3.15595	9.97998
9.47	89.6809	3.07734	9.73139	9.97	99.4009	3.15753	9.98499
9.48	89.8704	3.07896	9.73653	9.98	99.6004	3.15911	9.98999
9.49	90.0601	3.08058	9.74166	9.99	99.8001	3.16070	9.99500
				10.00	100.000	3.16228	10.0000

GLOSSARY OF SYMBOLS AND ABBREVIATIONS

SYMBOL	MEANING	FRAME*
A.D.	Average Deviation	6–24
b_{xy}	Regression coefficient for variable X on variable Y	20–31
b_{yx}	Regression coefficient for variable Y on variable X	20–15
D	Difference between two scores or ranks	4–42, 21–23
d	Deviation of differences between a pair of scores from the mean difference	21–32
df	Degrees of freedom	10–24
E	Expected frequency (in chi square test)	23–9
F	Analysis of variance technique	17–4
f	Frequency of scores	1–15
f_w	Frequencies of scores within a class interval	2–21

*"6–24" means set 6, frame 24. Frame and set designations refer the student to the frame in which the discussion of the symbol begins, not necessarily the first one in which it appears. It may occasionally be necessary to read several frames in order to obtain an adequate definition or abbreviation.

SYMBOL	MEANING	FRAME*
i	Number of values encompassed by a class interval	1–35
$\ell\ell$	Lower limits of a class interval	1–46
Mdn	Median	2–12
MS	Mean square (in analysis of variance)	16–13
N	Number of raw scores or ranks	1–20
O	Observed frequency (in chi square test)	23–9
P	Probability	8–2
Q	Semi-interquartile range	5–8
Q_1	Quartile designation (1st quartile)	4–35
r	Pearson product-moment correlation coefficient	19–3
s'	Sample standard deviation	7–19
s'^2	Sample variance	7–2
s	Estimate of population standard deviation	8–38
s^2	Estimate of population variance	10–10
$s_{\overline{X}}$	Estimate of the standard error of the mean	11–3
s_D	Estimate of the standard deviation of differences between pairs of raw scores	21–33
$s_{\overline{X}_1 - \overline{X}_2}$	Estimate of the standard error of the difference between means	14–28
SS_b	Sum of squares between groups (in analysis of variance)	16–13
SS_w	Sum of squares within groups (in analysis of variance)	16–13

SYMBOL	MEANING	FRAME
SS_t	Sum of squares of total groups (in analysis of variance)	16–13
t	The distribution derived from differences between μ and $\overline{X}$ divided by $s_{\overline{X}}$	12–21
X	Raw score	1–16
$\overline{X}$	Sample mean score	8–36
$\tilde{X}$	Estimated raw score (in regression equation)	20–30
x	Deviation of a raw score from the mean	6–7
$|x|$	Absolute deviation score	6–17
$\tilde{Y}$	Estimated raw score (in regression equation)	20–14
z	Relative deviate	7–27
Σ	The sum of	2–18
μ	Population mean score	3–3
σ	Population standard deviation	6–35
$\sigma_{\overline{X}}$	Standard error of sample means	9–16
σ^2	Population variance	10–15
$\sigma_{\overline{X}_1 - \overline{X}_2}$	Standard error of the difference between sample means	13–24
ρ	Rank order correlation	22–19
X^2	Chi square test of significance for frequency data	23–11

INDEX *

*'6-24' means set 6, frame 24. Page numbers are indicated in parentheses.